Revise AS Biology for AQA Specification B

Graham Read and Ray Skwierczynski

Contents

Introduction – How to use this revision guide

This revision guide is for the AQA Biology AS course. It is divided into modules to match Specification B. You may be taking a test at the end of each module, or you may take all of the tests at the end of the course. The content is exactly the same.

Each module begins with an **introduction**, which summarises the content. It also reminds you of the topics from your GCSE course which the module draws on.

The content of each module is presented in **blocks**, to help you divide up your study into manageable chunks. Each block is dealt with in several spreads. These do the following:

- summarise the **content**;
- indicate **points to note**;
- include **worked examples** of calculations;
- include **diagrams** of the sort you might need to reproduce in tests;
- provide **quick check** questions to help you test your understanding.

At the end of each module, there are longer **end-of-module questions** similar in style to those you will encounter in tests. **Answers** to all questions are provided at the end of the book.

You need to understand the **scheme of assessment** for your course. This is summarised on page v overleaf. At the end of the book, you will find some **exam tips** to help prepare you for the examinable component of the course.

AQA AS Biology – Assessment

You can take an AS as a qualification on its own or as the first part of an Advanced Level (A level) qualification. An AS forms 50% of an Advanced Level.

The AS offered by AQA Specification (Syllabus) B (S416) has four compulsory Units (Modules): Core Principles, Genes and Genetic Engineering, Physiology and Transport, and Coursework (Practical). Coursework is dealt with by your school or college. This book is designed to prepare you for the exams for the other Units.

About the tests

To get an AS, you will have to take a $1\frac{1}{4}$ hour exam for each Module/Unit: Core Principles, Genes and Genetic Engineering, and Physiology and Transport. All the questions have to be answered in each exam. Most of the marks will be for remembering and being able to explain the information in each Module. What you need to know is given in this book, at the level you need to know it. If you know more than that, it will certainly not harm you but it will not be needed to pass an exam. There is no synoptic element in these exams (that comes at Advanced Level). This means that you do not have to know the information in Core Principles to take the exam in Physiology and Transport, for example.

The table shows an outline of the AQA Specification B AS.

AQA Specification B – AS Examination
Core Principles – Assessment Unit 1 $1\frac{1}{4}$ **hour exam**   **30% of the AS**
Genes and Genetic Engineering – Assessment Unit 2 $1\frac{1}{4}$ **hour exam**   **30% of the AS**
Physiology and Transport – Assessment Unit 3(a) $1\frac{1}{4}$ **hour exam**   **25% of the AS**
Coursework – Assessment Unit 3(b) [Not covered in this book]   **15% of the AS**

Module 1: Core Principles

This module is broken down into six topics: Biological molecules; Cells; Cell transport; Organisms exchange materials with their environment; Enzymes; and Digestion. Cell transport and Organisms exchange material with their environment are closely related and are grouped together below.

Biological molecules

- Organisms are made up of a vast number of types of carbon-based biological molecules. Carbon can form bonds with other carbon atoms to make molecules with chains or rings of carbon atoms of different sizes, leading to an enormous range of possible molecules.
- Relatively small biological molecules called monomers can join together to form very large molecules called polymers, giving an even greater range of possible molecules. Other elements or chemical groups are attached to the basic carbon chains or rings, giving even more possibilities.
- Biological molecules are grouped into types of molecules with similar properties: carbohydrates, proteins, lipids and nucleic acids (dealt with in Module 2) are particularly important.

Cells

- Life exists as cells. There are two main types of cells: prokaryotic and eukaryotic.
- Eukaryotic cells are larger and have membrane-bound organelles.
- Cells of multicellular organisms are eukaryotic and differentiated; their size, shape, organelles and chemistry are adapted to a specific function.
- Differentiated cells form tissues and these are part of organs, which may be part of a system.

Cell transport and Exchange materials

- The chemical environment inside cells is different from the environment outside in terms of the types of molecules present and the concentrations of ions and molecules. These differences are essential for the survival of the cell, and exist only because the cell surface membrane controls what enters and leaves the cell: the membrane is differentially/selectively permeable.
- Cells have to exchange substances with their environment and this involves diffusion, osmosis and active transport.
- Small organisms (including single-celled ones) have a large surface area compared to their volume, and can exchange substances directly with their environment by diffusion, osmosis and active transport across cell membranes.
- In large organisms, such as humans and flowering plants, most cells are a long way from the external environment, and the body surface is covered by a waterproof layer to prevent drying out (desiccation), which is a barrier to exchange. These organisms have special exchange surfaces.

Enzymes

- Enzymes are proteins which are biological catalysts. They allow specific reactions in cells to take place quickly under the environmental conditions found in cells.

Digestion

- Enzymes are required for digestion, the process that breaks down large biological molecules into small, soluble molecules that can cross cell membranes to enter cells and organisms.
- Simple organisms secrete digestive enzymes onto their food.
- Humans ingest food into the gut, which is adapted for efficient digestion and absorption.

Carbohydrates – mono and disaccharides

Carbohydrates contain the elements carbon, hydrogen and oxygen. The hydrogen and oxygen are present in the ratio of 2 hydrogen : 1 oxygen.

Carbohydrates can be can be classified into three groups: **monosaccharides**, **disaccharides** and **polysaccharides**. Monosaccharides and disaccharides are small, soluble molecules that are easy to transport and sweet to taste.

Monosaccharides (single sugars)

Monosaccharides are the basic molecular units (**monomers**) of carbohydrates. They are mainly used in respiration and in growth during the formation of larger carbohydrates. They:

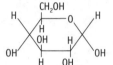

Structure of glucose

- include **glucose** and **fructose** which have the same formula, $C_6H_{12}O_6$ (**hexoses**), but different structures (**isomers**);

- are **reducing sugars**, so give a positive (+ve) Benedict's test result (**brick red** precipitate/colour).

Disaccharides (double sugars)

A disaccharide is formed when two monosaccharides are joined together by a **condensation reaction**.

Examples of disaccharides are **sucrose** and **maltose**.

Glucose	**+**	**Glucose**	→	**Maltose**	**+**	**Water**
Glucose	**+**	**Fructose**	→	**Sucrose**	**+**	**Water**
$C_6H_{12}O_6$	**+**	$C_6H_{12}O_6$		$C_{12}H_{22}O_{11}$	**+**	H_2O
Monosaccharide		**Monosaccharide**		**Disaccharide**		

> All the large biological molecules you study are formed from smaller molecules by condensation reactions – where water is formed. They can all be broken down by hydrolysis reactions, using water.
>
> ✓ *Quick check 1, 2*

A disaccharide can be broken down into its monosaccharides by a **hydrolysis reaction**. Hydrolysis is the use of water (hydro) in the breakdown (lysis) of a larger molecule into smaller molecules. Disaccharides can be hydrolysed by boiling with acid, e.g. dilute HCl, or by an enzyme at its optimum temperature.

glucose   glucose

H_2O hydrolysis H_2O condensation

maltose

Formation and hydrolysis of maltose

Example: Hydrolysis of sucrose

Sucrose	+	Water	$\xrightarrow{\text{enzyme sucrase}}$	Glucose	+	Fructose
$C_{12}H_{22}O_{11}$	+	H_2O		$C_6H_{12}O_6$		$C_6H_{12}O_6$
Disaccharide				Monosaccharides		

Benedict's test for reducing sugars

- A small amount of the food sample/solution being tested for reducing sugar is placed in a test tube with 2 cm^3 of Benedict's solution.

- This is heated in a boiling water bath for 5 minutes.

- A **brick red** precipitate/colour is a **+ve** result.

- Glucose, fructose and maltose give +ve results.

- If the Benedict's solution remains **blue**, no reducing sugar is present.

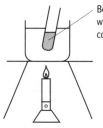

Benedict's solution turns brick red when heated with solution containing reducing sugar

Benedict's test

✓ Quick check 3

Test for a non-reducing sugar (sucrose)

Sucrose is a non-reducing sugar and can be identified by the following method.

- Carry out Benedict's test on a sample to confirm a negative result.

- Hydrolyse another sample by heating it with dilute acid, e.g. HCl, or by using the enzyme **sucrase** (invertase) at its optimum temperature.

- When cooled, add dilute sodium hydroxide solution (NaOH) to neutralise the acid.

- Add Benedict's solution and heat in a water bath for 5 minutes.

- A +ve brick red colour indicates a **non-reducing sugar (sucrose)** was present in the sample.

? Quick check questions

1 Give two differences between glucose and sucrose.

2 Explain how a disaccharide is formed from two monosaccharides.

3 Describe how you would test a sample of apple juice for the presence of reducing sugar.

Carbohydrates – polysaccharides

Polysaccharides are large polymers made from monosaccharide monomers. The polysaccharides cellulose, starch and glycogen are polymers of glucose, and are formed by joining many glucose monomers (molecules) by condensation reactions.

- Polysaccharides differ in the number and arrangement of the glucose molecules they contain.
- They function as storage or structural molecules.
- They are not sweet to taste and are relatively insoluble in water.

> ▶ Polysaccharides are **non-reducing** – giving a negative result in Benedict's test.

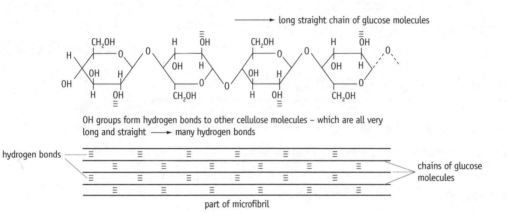

Structure of cellulose

Cellulose

> ▶ Cellulose is hydrolysed by the enzyme **cellulase**.

Cellulose is found in the cell wall of plants, which provides rigidity and shape to the cell.

- Cellulose has thousands of glucose molecules joined together by condensation reactions to form a **long, straight chain**.
- **Hydroxyl** (OH) groups on each chain form weak **hydrogen bonds** with hydroxyl groups of other chains, producing a **microfibril**.
- The long, straight chains allow cellulose molecules to lie parallel to each other and form very many hydrogen bonds which together hold the molecules very firmly together – making microfibrils very strong.
- Microfibrils are grouped into larger bundles known as macrofibrils.
- Macrofibrils in one layer are orientated in the same direction.
- Macrofibrils in successive layers are orientated in a different direction.
- Macrofibrils of these different layers are interwoven and embedded in a matrix – making the cell wall **rigid**.
- The cellulose cell wall is usually **fully permeable** due to small channels between the different layers of macrofibrils.

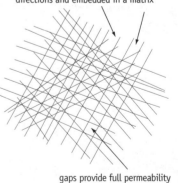

layers of macrofibrils orientated in different directions and embedded in a matrix

gaps provide full permeability

Arrangement of macrofibrils

✓ *Quick check 1*

Starch

This is the storage carbohydrate found in plants, consisting of long, **branched** chains of glucose molecules. Starch is stored in starch grains (amyloplasts) in the cytoplasm. Starch is ideally suited to its function as a storage compound because:

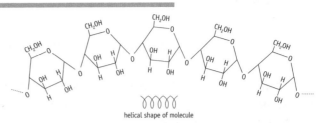

helical shape of molecule

Structure of starch

- it is **insoluble** and therefore does not affect the **water potential** of the cell and water movement due to osmosis (it is osmotically inactive);

- the molecule has a **helical** shape, forming a **compact store**;

- it contains a large number of **glucose molecules** providing a large supply of **respiratory substrate**;

- it is too large to cross the cell membrane and so stays where it is formed.

> ◗ Foods obtained from plants usually test positive for starch. Starch is not stored in animal/human cells.

Example: starch is hydrolysed by the enzyme amylase to produce the disaccharide maltose.

Starch + Water	Amylase	Maltose
Polysaccharide	Hydrolysis →	Disaccharide

✓ *Quick check 2*

Starch can be detected in a sample by using the iodine test. This involves:

- adding two or three drops of potassium iodide solution;

- if starch is present a blue/black colour is produced;

- if no starch is present the iodide solution remains orange/yellow.

Glycogen

This is the storage carbohydrate found in the cytoplasm of animal cells – sometimes referred to as 'animal starch' due to similarities in structure and function between the two molecules. Glycogen:

- is a polymer of glucose, similar to starch but with more branches;

- has the same storage advantages as starch;

- is stored in large amounts in liver and muscle tissues.

✓ *Quick check 3*

? Quick check questions

1 Explain how the structure of the cellulose molecule leads to the rigidity of a plant cell wall.

2 Give two features of a starch molecule that are related to its function as a storage compound.

3 Name one tissue that contains a large amount of glycogen.

Lipids

Lipids contain the elements carbon, hydrogen and oxygen. They possess a lower proportion of oxygen and a higher proportion of hydrogen compared to carbohydrates. They are used in respiration and as structural and storage molecules. Lipids:

- are insoluble in water, but are soluble in a range of organic solvents, e.g. alcohol;

- being insoluble, do not affect the **water potential** of the cell and water movement due to osmosis (they are osmotically inactive);

- are ideal storage compounds yielding twice as much energy per gram as carbohydrates when fully oxidised (respired), and consist of fats (solids) and oils (liquids) at room temperature.

Triglycerides are one type of lipid, formed by joining three **fatty acids** to one **glycerol** molecule by **condensation** reactions.

- The general formula of a fatty acid is **R-COOH.**

- All fatty acids have **COOH** which is a carboxylic acid group.

- Different fatty acids have different R groups – representing long hydrocarbon chains which differ in the number of carbon atoms they contain and in whether they are **unsaturated** (contain double bonds) or **saturated** (no double bonds).

- The three fatty acids in a lipid may be the same (simple lipid) or different (mixed lipid).

Hydrolysis of lipids

Lipids can be hydrolysed into fatty acids and glycerol by:

- heating with acid or alkali;

- using the enzyme **lipase** at its optimum temperature.

Structure of a glycerol molecule

✓ *Quick check 1*

▶ Make sure you know the structures of biological molecules, but also learn and try to understand how these structures are related to the functions of the molecules.

✓ *Quick check 2*

Formation and hydrolysis of a lipid molecule

Emulsion test for lipids

Lipid can be detected in a sample using the emulsion test. This involves:

- a small amount of the sample being placed into a test tube with 2 cm^3 of **ethanol**;

- the mixture being shaken so that the fat dissolves;

- adding this to water in another test tube and shaking the contents;
- a **white or cloudy emulsion** of fat droplets indicates fat is present.

Phospholipids

Phospholipids are lipids containing a phosphate group. In the commonest type of phospholipid, a phosphate group replaces one of the fatty acid molecules.

- A phospholipid molecule consists of one glycerol, two fatty acids and a phosphate group joined by **condensation** reactions.
- The phospholipid molecule has a polar **hydrophilic** head (attracts water) containing the phosphate group, and a non-polar **hydrophobic** tail (repels water) consisting of the long fatty acid chains.

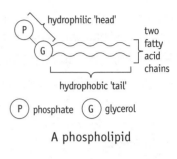

A phospholipid

✓ *Quick check 3*

Phospholipid layers

- Phospholipids on the surface of water form a single layer as the hydrophilic heads are attracted to the water and the hydrophobic tails are repelled by water and project outwards – see (a) below.
- In the cell membrane a **bilayer** (two layers) of phospholipids is present as the hydrophobic tails are attracted to each other and away from water inside and outside the cell – see (b) below.

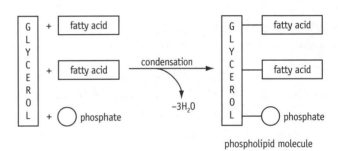

Formation of a phospholipid

(a)

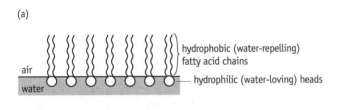

(b)

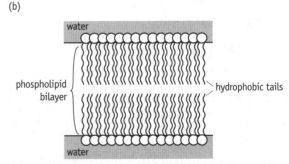

(a) A single phospholipid layer; (b) a bilayer

Quick check questions

1 What is a saturated fatty acid?

2 Give two ways by which lipids could be hydrolysed.

3 How does a phospholipid molecule differ from a triglyceride?

Proteins

Proteins contain carbon, hydrogen, oxygen, nitrogen (and sometimes sulphur). Amino acids are the monomers from which proteins (polymers) are formed.

- There are 20 different, commonly occurring amino acids in living organisms.
- All amino acids have an amino group and a carboxylic acid group but differ in the structure of their R groups.
- Amino acids are joined together by **condensation** reactions (peptide bonds are formed).
- Two amino acids are joined together to form a **dipeptide**.
- Many amino acids are joined together to form a **polypeptide**.
- A protein may consist of one or more polypeptides.

Generalised structure of an amino acid

✓ *Quick check 1, 2*

Hydrolysis of proteins

Proteins can be hydrolysed by heating with acid or by enzymes (**proteases**). These enzymes include:

- **endopeptidases**, which hydrolyse bonds between amino acids in proteins to produce smaller polypeptides;
- **exopeptidases**, which hydrolyse bonds at the ends of polypeptide chains producing dipeptides and single amino acids.

Biuret test for proteins

Protein can be detected in a sample by:

- adding it to a test tube containing 2 cm^3 of dilute **sodium hydroxide** solution;
- dilute **copper sulphate** solution is then added drop by drop;
- a **purple, lilac or mauve** colour indicates **protein** is present;
- if the solution remains blue, no protein is present.

Formation and hydrolysis of proteins

Structure of proteins

Proteins vary in the number, type and arrangement of amino acids they contain. This produces a vast number of different protein molecules.

- **Primary structure** is the sequence of amino acids in the polypeptide chain of a particular protein.
- Specific interactions between different parts of the chain produce the final **3-dimensional** shape of the protein.

Primary structure of a protein

✓ *Quick check 3*

- **Secondary structure** is the folding or coiling of the polypeptide due to hydrogen bonding between amino acids in the chain.
- Secondary structures include the α-**helix** and the β-**pleated sheet**.
- **Tertiary structure** is the 3-dimensional shape of the protein due to folding and coiling of the secondary structure caused by specific **hydrogen bonds**, **ionic bonds** and **disulphide bridges** between parts of the chain.

Proteins can be classified according to their structure.

- A **fibrous protein's** primary structure produces very large (insoluble) and long polypeptide chains with a simple tertiary structure.
- These wrap around each other to form strong, insoluble fibres, e.g. collagen in connective tissue.
- A **globular protein's** primary structure produces a highly folded polypeptide chain with a very complex and specific tertiary structure.
- This allows it to recognise and bind specifically to another molecule (or ion), e.g. enzymes, antibodies and some hormones.
- Their amino acid composition, size and folding make them soluble and easy to transport.

> The tertiary structure or 3-dimensional shape of a protein is almost always the key to any property of the protein, and to any question!

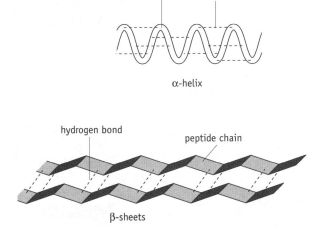

α-helix

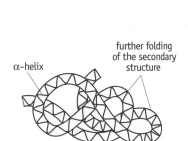

Tertiary structure of a protein

Secondary structures of proteins

Denaturation is a change in the tertiary structure of a protein – making it no longer functional.

- It is caused by temperatures above the optimum or extreme changes in pH.
- These conditions break hydrogen and ionic bonds.

✓ *Quick check 4*

? *Quick check questions*

1 Give one element present in all protein molecules which is not present in carbohydrates.

2 Give the generalised structural formula of an amino acid.

3 Explain what is meant by the primary structure of a protein.

4 Explain what happens to protein molecules when heated to high temperatures.

Chromatography

Chromatography is a method of separating and identifying compounds present in a mixture. You need to know about paper chromatography. Paper chromatography separates the components of a mixture by their different solubilities in a particular solvent.

The apparatus for paper chromatography is set up as on the right.

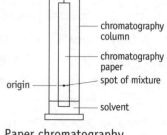

Paper chromatography

- The solvent moves up the paper by capillary action carrying the components of the mixture.

- The more soluble a component is in the solvent, the further it is carried up the paper.

- When the solvent is near to the top of the paper, the paper is removed and a line drawn to mark the **solvent front**.

- The chromatogram is air-dried and often treated with a locating agent to stain any colourless components which then appear as spots.

- The distance of each spot from the origin is measured.

Identification of compounds

> Insoluble compounds e.g. starch remain at the origin when carrying out chromatography.

Separated compounds can be identified by their R_f values. The R_f value of a compound is constant in a specific solvent, enabling the compound to be identified from standard R_f tables. Alternatively, the position of the spots can be compared with the position of known compounds (markers) which have been placed alongside the mixture or on a separate chromatogram.

> Chromatography can be used to separate and identify the different sugars present in fruit juice.

$$R_f \text{ value} = \frac{\text{distance moved by compound from origin}}{\text{distance moved by solvent from origin}}$$

✓ *Quick check 1, 2*

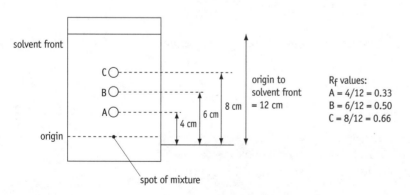

origin to solvent front = 12 cm

R_f values:
A = 4/12 = 0.33
B = 6/12 = 0.50
C = 8/12 = 0.66

Determination of R_f values

Two-dimensional chromatography

Some compounds have similar R_f values in a particular solvent so their spots overlap. Two-dimensional chromatography uses a second solvent to separate these compounds. The diagram below shows the method involved.

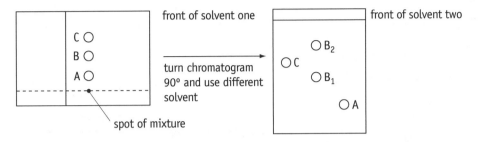

Two-dimensional chromatography

The second solvent front is marked and the R_f values of the spots are calculated. In the diagram, compounds B1 and B2 have similar R_f values in the first solvent but different R_f values in the second solvent.

✓ **Quick check 3**

Uses of chromatography

Paper chromatography can be used in a laboratory to separate and identify:

- the different pigments of chlorophyll;
- the amino acids present in a protein after it has been completely hydrolysed;
- monosaccharides and disaccharides.

? Quick check questions

1 Give the formula used to calculate R_f values.

2 Explain why locating agents are sometimes required.

3 Explain fully how you could use chromatography to ensure the separation and identification of two similar compounds.

Water

Water is the most abundant compound on earth and makes up between 60 and 95% of all living organisms. A water molecule consists of two hydrogen atoms covalently bonded to an oxygen atom. Weak hydrogen bonds form between water molecules, and these attractive forces are responsible for many of the properties of water.

hydrogen bonds between water molecules

Hydrogen bonding between water molecules

Role of water as a solvent

Water is a solvent for numerous biochemical molecules, giving solutions and enabling:

- transport of nutrients, e.g. glucose and amino acids in blood, and sucrose in phloem;

- removal of excretory products, e.g. ammonia, urea;

- secretion of substances, e.g. hormones, digestive juices.

> Insoluble storage compounds such as starch and glycogen are converted into soluble products to enable their transport to other parts of an organism.

Role of water in metabolic reactions

✓ *Quick check 1, 2*

The majority of essential metabolic reactions take place in solution in water. Water is a raw material or a product of many metabolic reactions.

- Hydrolysis involves the addition of water (hydro) in the breakdown (lysis) of large biological molecules into their monomers/sub-units, e.g. proteins into amino acids. (Water is released during condensation reactions.)

- Water produced as a metabolic product of respiration is essential for organisms, especially those living in dry habitats.

- Water is needed for photosynthesis.

Role of water in support

Water is not easily compressed and has an important role in support in plants and animals.

- The uptake of water by plant cells creates a pressure against the rigid cell wall.
- This turgor pressure helps non-woody (herbaceous) plants to remain upright.
- Water provides buoyancy for aquatic organisms, e.g. whales.
- Water has a high surface tension and water molecules have cohesive forces holding them together, due to hydrogen bonding between water molecules.
- These properties allow aquatic insects to walk on the surface of water, and water to be pulled through xylem in plants.

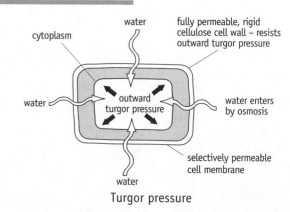

cytoplasm

water

fully permeable, rigid cellulose cell wall – resists outward turgor pressure

water

outward turgor pressure

water enters by osmosis

selectively permeable cell membrane

water

Turgor pressure

Role of water in temperature regulation

Water has a **high specific heat capacity** which means it absorbs a lot of heat energy for its temperature to rise, and loses a lot to cool. This helps to:

- reduce temperature fluctuations in organisms – especially large ones;
- minimise increases in temperature in cells as a result of biochemical reactions;
- reduce fluctuations in temperature in aquatic habitats.

A lot of heat is needed to turn water into vapour (it has a high latent heat of vaporisation).

- This helps some animals to maintain a constant body temperature as a high amount of heat energy is removed from the body to evaporate sweat or during panting.
- In plants evaporation of water from leaves has a cooling effect.

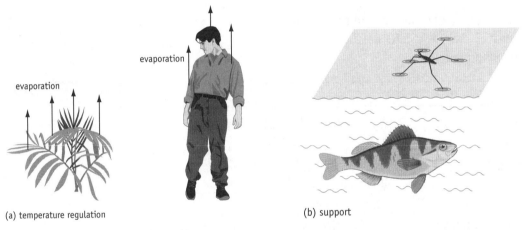

evaporation

evaporation

(a) temperature regulation

(b) support

Some functions of water

At 4°C water is at its maximum density and becomes less dense as it freezes.

- Water is denser as a liquid than as a solid.
- Thus cold water forms ice on the upper surface, insulating the aquatic organisms below.
- Water must lose a relatively large amount of heat energy to freeze, making the formation of ice crystals in cells less likely.

✓ *Quick check 3*

? *Quick check questions*

1. Give two examples of how water functions as a solvent in living organisms.
2. Name the bonds responsible for the attractive forces between water molecules.
3. Water has a high specific heat capacity. Explain what this means and why this is important to living organisms.

Cells

The cell is the basic unit of living organisms. Cells only come from existing cells by cell division. Cells and organisms can be divided into two main groups: prokaryotes and eukaryotes.

Comparison of prokaryotes and eukaryotes

Bacteria are prokaryotes and are very small, single-celled organisms with **no nucleus** and **no other membrane-bound organelles**.

- A typical bacterial cell has a cell wall, cell membrane, genetic material (DNA), small (70S) ribosomes, and cytoplasm.

- A capsule, plasmids and flagellae may be present.

Eukaryotes form the vast majority of types of living organisms, including plant and animal. **Eukaryotic cells** have a nucleus and many other membrane-bound organelles.

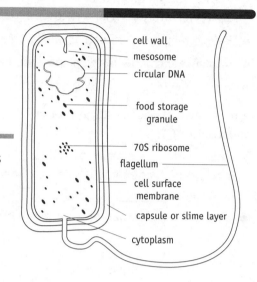

Typical bacterial cell as seen with an electron microscope

✓ **Quick check 1**

Prokaryotic	Eukaryotic
DNA is circular, not in a nucleus	DNA is linear, within a nucleus
Diameter of cell 0.5–10 μm	Diameter of cell 10–100 μm
Smaller, lighter 70S ribosomes	Larger, heavier 80S ribosomes
No mitochondria present	Mitochondria present
No Golgi body	Golgi body present
Flagella (when present) simple, lacking microtubules	Flagella (when present) have microtubules

Eukaryotic plant cells also have:
- cellulose cell walls providing support and shape to the cell;
- starch grains;
- chloroplasts (in photosynthetic cells), containing the pigment chlorophyll;
- a large vacuole containing soluble sugars, salts and sometimes pigment.

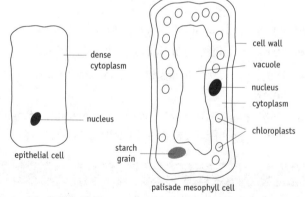

An epithelial cell from the small intestine and a palisade mesophyll cell as seen with a light microscope

Electron microscopy

The electron microscope is used to investigate the fine structure (**ultrastructure**) of a cell. It uses a beam of electrons focused by electromagnets, as opposed to the focusing of light rays by lenses in light microscopy.

Advantages of electron microscopy

- Electrons have a shorter wavelength than light, giving **greater resolution** (the ability to distinguish between two close objects). Light microscopy has a maximum resolution of around 0.2 μm whereas with the electron microscope it is 200 nm.

❱ Don't confuse resolution and magnification. You could get a magnification of 500 000 with a light microscope but the image would be very blurred due to poor resolution.

- The **maximum useful magnification** is higher than with light microscopy.

Disadvantages of electron microscopy

- As a vacuum is required, living specimens cannot be seen.
- Preparation and staining techniques can alter/damage cells.
- Expensive to buy and expert training is required in their use.

✓ *Quick check 2*

Differential centrifugation

> Isotonic refers to solutions with the same water potential but not necessarily the same composition.

✓ *Quick check 3, 4*

Centrifugation separates structures of different mass or density. Differential centrifugation involves centrifuging at different speeds to separate the different organelles in a cell.

- Cells are broken open (**homogenised**) by grinding tissue in ice-cold, **isotonic** buffer solution, using a blender.

- An isotonic solution prevents the net osmotic movement of water in or out of organelles, which might cause them to burst or shrivel.

- A low temperature prevents the action of enzymes that cause self-digestion (autolysis) of organelles.

- The suspension is poured into a tube and spun in a centrifuge at low speed and cell debris, e.g. cell walls of plant cells, forms a **sediment** (**pellet**) at the bottom.

- The **supernatant** (liquid above the sediment) is spun at a higher speed (generating greater force).

- The heaviest organelles, the **nuclei**, are forced to the bottom and form a sediment, whilst lighter organelles remain suspended in the supernatant.

- The procedure is repeated, increasing the speed of centrifugation each time giving a series of pellets containing organelles of decreasing mass/density.

- The organelles are usually isolated in the order: nuclei, chloroplasts, mitochondria, endoplasmic reticulum, lysosomes, and finally ribosomes.

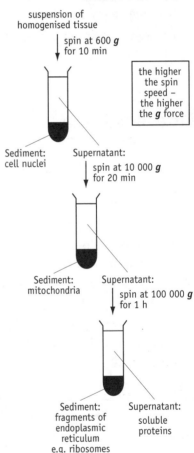

Differential centrifugation of cell organelles

? ## Quick check questions

1 Give three ways in which prokaryotes and eukaryotes differ in structure.

2 Give one advantage and one disadvantage of using an electron microscope.

3 Explain why (i) a low temperature and (ii) an isotonic solution are required during centrifugation.

4 Which organelle will mainly be present in the first sediment during differential centrifugation?

Cell structure – organelles

Organelles in eukaryotic cells include the nucleus, mitochondria, Golgi body, rough and smooth endoplasmic reticulum, chloroplasts and ribosomes.

You must be able to recognise the different cell organelles as shown below.

Nucleus

The nucleus contains the genetic material, DNA, and:

- is bound by a double membrane, the **nuclear envelope**, which has **nuclear pores** allowing communication with the cytoplasm;

- contains **chromatin** (DNA and protein) and **nucleoli** (in a dividing cell the chromatin is seen as **chromosomes**);

- controls protein synthesis, cell division and production of ribosomes.

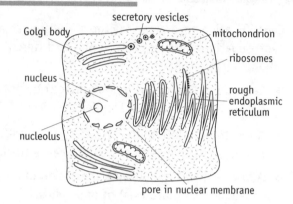

A generalised animal cell seen with an electron microscope

Mitochondria

Mitochondria are involved in **aerobic respiration** that produces ATP. They are often rod-shaped and between 1 and 10 μm in length. Mitochondria:

- are bounded by two membranes forming an envelope around the **matrix** – the site of enzymes involved in respiration and also containing DNA and ribosomes;

- have an inter-membrane space between the outer and inner membranes;

- have **cristae** – folds of the inner membrane which provide a large surface area for enzymes associated with ATP production needing oxygen (oxidative phosphorylation).

A mitochondrion

Cells requiring large amounts of ATP have numerous mitochondria, e.g. cells involved in active transport, and muscle cells.

Ribosomes

Ribosomes are very small organelles (20 nm diameter), made of one large and one small subunit, each made up of protein and RNA. Ribosomes:

- can be present in the cytoplasm singly, in a chain (polysomes) or attached to the rough endoplasmic reticulum;

- are used in **protein synthesis** – where condensation reactions join amino acids together.

Endoplasmic reticulum

Consists of flattened membrane sacs called **cisternae**.

- The surface of the **rough endoplasmic reticulum** has ribosomes that produce proteins into the cisternae.
- The **smooth endoplasmic reticulum** lacks ribosomes and is involved in the transport of substances around the cell, and the production of lipids.

Golgi body

A Golgi body consists of a stack of flattened, membrane-bound sacs – **cisternae**.

- The cisternae are continually being formed at one end and pinched off as **Golgi vesicles** at the other.
- Secretory cells have large Golgi bodies.

The Golgi body and vesicles:

- produce glycoproteins, e.g. mucin;
- package and secrete enzymes;
- form cell walls in plant cells;
- form lysosomes.

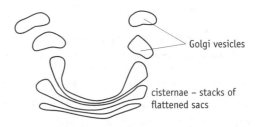

A Golgi body

▶ It is quite easy to learn to label the parts of organelles – make sure you do!

Chloroplasts

Chloroplasts are only found in photosynthetic plant cells.

- They are flattened discs, 3–10 μm in diameter, bounded by two membranes.
- Inside is a membrane system of many flattened sacs called **thylakoids**, which in places form stacks called **grana**.
- Grana contain chlorophyll molecules which **absorb light** for **photosynthesis**.
- The membrane system is surrounded by the **stroma** – the site of the enzymes used to make sugars and starch during photosynthesis.

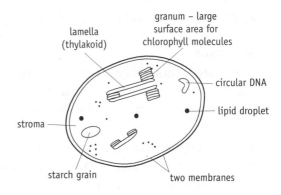

Structure of a chloroplast

✓ *Quick check 1, 2*

? ## Quick check questions

1 Give two ways in which the structure of a mitochondrion and a chloroplast are similar.
2 Give one function of the following organelles: (i) ribosomes; (ii) smooth endoplasmic reticulum; (iii) Golgi body.

Cell membranes and cell differentiation

The entry and exit of substances into and out of cells is controlled by the **cell membrane** or **plasma membrane**, which surrounds the cytoplasm of a cell. The structure of the plasma membrane is described by the **fluid-mosaic model**, and is basically the same for membranes around cell organelles.

The plasma membrane

The cell membrane consists of protein and phospholipid. The phospholipid molecules form a bilayer but are constantly moving about, giving a **fluid** structure. The protein molecules are unevenly distributed throughout the membrane, forming a **mosaic**. The **selective permeability** of the cell membrane is related to the type and distribution of protein and phospholipid molecules present in the membrane. In the cell membrane:

- a double layer of phospholipid molecules – the **phospholipid bilayer** – has protein molecules partially embedded or penetrating completely through;
- the **hydrophobic tails** (fatty acid chains) of the phospholipid molecules in the bilayer are attracted towards each other;
- the **hydrophilic heads** of the bilayer are orientated either inwards towards the cytoplasm or outwards toward the watery extra-cellular fluid;
- the presence of cholesterol decreases permeability and increases stability of the membrane;
- the **phospholipid bilayer** allows lipid-soluble molecules to pass through but restricts passage of ions and polar molecules;
- ions and polar molecules can pass through **channel proteins** (pores) that span the membrane;
- other **protein** molecules act as **carriers** – aiding the transport of ions and polar molecules, e.g. glucose, by facilitated diffusion and active transport;
- other protein molecules act as specific **receptors** for hormones, e.g. insulin, which attach allowing the cell to respond.

> The proteins acting as channels, carriers and receptors are not all the same things.

✓ *Quick check 1, 2*

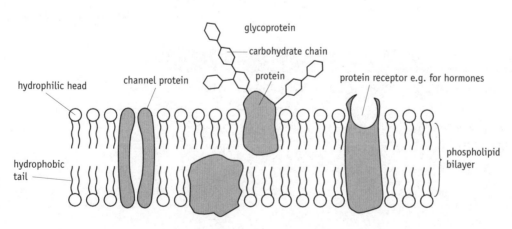

Structure of the cell membrane

Cell differentiation

As multicellular organisms evolved, specialised cells developed to perform specific functions. The development of specialised cells, e.g. neurones and muscle cells, is known as cell differentiation. These cells are usually grouped together to form **tissues** and **organs.**

- **Tissues** are collections of similar cells that perform a specific physiological function, e.g. epithelial, nervous or muscular tissue in animals, mesophyll, phloem or xylem tissues in plants.

- An **organ** is a structure consisting of different tissues, which has a specific physiological function, e.g. the stomach, which has a role in the digestion of food.

- Several organs combine to form a **system**, e.g. the stomach, liver and pancreas all form part of the digestive system.

✓ *Quick check 3*

Epithelial cells

Epithelial cells in the small intestine are specifically adapted for the absorption of digested food products by having:
- the cell membrane facing the gut contents folded into **microvilli** – providing a large surface area for absorption;
- numerous mitochondria – providing **ATP** for the active uptake of digested food molecules.

▶ These two types of cell are the only ones specifically mentioned by the syllabus.

Palisade mesophyll cells

Palisade mesophyll cells in the leaf are adapted for photosynthesis by having:
- many chloroplasts to absorb light;
- a thin cell wall – short diffusion pathway for carbon dioxide;
- an elongated shape – allowing many cells to be packed closely side-by-side and to absorb as much light as possible.

✓ *Quick check 4*

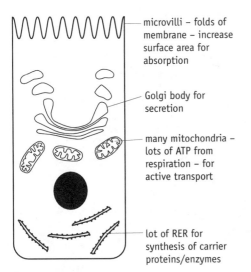
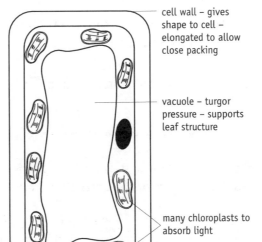

microvilli – folds of membrane – increase surface area for absorption

Golgi body for secretion

many mitochondria – lots of ATP from respiration – for active transport

lot of RER for synthesis of carrier proteins/enzymes

cell wall – gives shape to cell – elongated to allow close packing

vacuole – turgor pressure – supports leaf structure

many chloroplasts to absorb light

Adaptations of an epithelial cell from the small intestine (left) and a palisade mesophyll cell (right)

? *Quick check questions*

1 Explain how lipid-soluble molecules are able to pass rapidly into a cell.

2 Give the function of: (i) protein carriers; (ii) receptors in the cell membrane.

3 Explain what is meant by: (i) a tissue; (ii) an organ.

4 Describe two ways in which an epithelial cell from the small intestine is adapted for its function.

Transport across membranes

Movement of substances into and out of cells through the plasma membrane can occur by **diffusion**, **facilitated diffusion**, **osmosis** and **active transport.**

Diffusion

Diffusion is a passive process – it does not require energy from respiration. Gaseous exchange occurs via diffusion.

> **Diffusion is the net movement of molecules from a higher concentration to a lower concentration until they are equally distributed.**

The rate of diffusion is increased by a greater concentration gradient, a large surface area (e.g. microvilli) and a short diffusion distance.

Facilitated diffusion

In this process channel proteins and carrier proteins transport ions and polar molecules across membranes. The specific tertiary structure of each type of protein means that it recognises, binds with and transports a specific substance.

- **Channel proteins** have a fixed shape whereas **carrier proteins** change shape as they move molecules, e.g. glucose, across the membrane.

- Facilitated diffusion does not require energy from respiration.

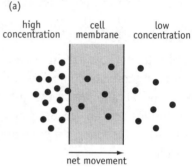

Diffusion (a) and facilitated diffusion (b)

Osmosis

> **Osmosis is the net movement of water molecules by diffusion from a dilute solution to a concentrated solution across a partially permeable membrane.**

In terms of water concentration – osmosis is the movement of water from a high concentration of water molecules to a low concentration of water molecules.

Water potential

In osmosis, water moves from a **higher (less negative)** water potential to a **lower (more negative)** water potential.

> **Water potential can be defined as the potential (tendency) of water molecules to leave a solution by osmosis.**

- The greater the concentration of water molecules in a cell or solution, the higher its water potential.

- **Pure water** has the **highest** water potential – **zero.**

Get used to using the terms used here to describe differences in water potentials. References to larger and smaller, or bigger and smaller water potentials will be marked wrong.

- A solution has a negative water potential – the more concentrated the solution, the lower (more negative) its water potential.

Consider solutions A and B shown on the right.

- Solution A is **hypotonic** to solution B – it has a lower concentration of solute molecules and a higher concentration of water molecules.

- Solution B is **hypertonic** to solution A – it has a higher concentration of solute molecules and a lower concentration of water molecules.

- **Net** water movement occurs from solution A to solution B, until the solutions are **isotonic** – equal concentration.

- At this point water movement will be equal in both directions.

- **Cell turgor** is essential in providing support in many plants.

- Ions (and sugars) in the vacuole of plant cells lower the water potential, and water enters by osmosis.

- The vacuole enlarges and the cell exerts an outward turgor pressure on the cell wall which resists this pressure and prevents a great expansion of the cell.

- When the cell has taken up as much water as it can, it is **turgid** – firm.

- Turgid cells push against each other, making tissues firm – providing support.

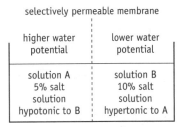

Osmosis

▶ In any question about movement of water in and out of cells it is very important to mention – the net diffusion of water, osmosis, selectively permeable membranes and water potentials.

✓ *Quick check 1, 2*

▶ You can think of turgor as being like inflating the inner tube of a bicycle tyre to make the tyre firm. The inner tube is inflated with air and pushes against the rigid tyre – which resists expansion.

Active transport

> **Active transport is the transport of molecules or ions across a membrane by carrier proteins against a concentration gradient.**

- It requires **energy – ATP** from respiration.

- Factors reducing the rate of respiration, reduce active transport:
  — lower temperature
  — lack of oxygen
  — metabolic and respiratory inhibitors.

- Active transport involves carrier proteins in the membrane.

- The hydrolysis of ATP releases the energy required for active transport.

- Cells involved in active transport have a large number of **mitochondria** to provide the ATP required via **aerobic respiration**.

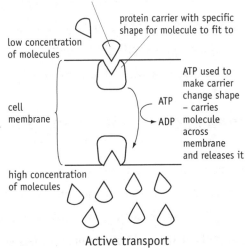

Active transport

✓ *Quick check 3, 4*

Quick check questions

1 What is the water potential value of pure water?

2 Explain the movement of water from the soil into a root hair cell.

3 Give two ways in which diffusion and active transport differ.

4 Give two ways by which you could reduce active uptake of ions by a cell.

Gaseous exchange in mammals

Gaseous exchange between an organism and its environment relies on diffusion. Very small organisms (e.g. amoebae) use diffusion through the outer surface. Larger organisms have developed internal gas exchange surfaces.

Surface area : volume ratio

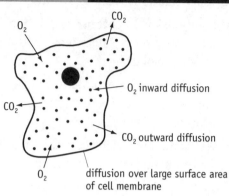

Gaseous exchange in an amoeba

- Single-celled organisms have a large surface area to volume ratio and short diffusion pathways to all parts of the cell.

- As the size of organisms increases, the surface area : volume ratio decreases.

Adaptations have evolved which maintain the adequate exchange of substances (e.g. nutrients, waste products, loss of heat).

- Changes in body shape increase the surface area for exchange.

- For example, flattening of the body in flatworms, and ears with a large surface area to lose heat in elephants.

- **Internal respiratory systems** – a large surface area relative to the volume of the organism.

> Remember – small organisms have a large surface to volume ratio, and vice versa for large organisms.

> ✓ *Quick check 1*

Respiratory surfaces

Respiratory surfaces generally have:

- large surface areas to increase the rate of diffusion;
- moist surfaces – for gases to dissolve in and pass through;
- thin surface layers (epithelia) – providing a short diffusion pathway;
- extensive blood circulation and a ventilation mechanism to maintain diffusion gradients.

Gaseous exchange in mammals

> You might get a question that talks about gas exchange in an organism you've never heard of. Don't panic – look for the basic properties of exchange surfaces – they must be there!

The gas exchange system in mammals is the lungs, which consist of:

- the **trachea** – supported by incomplete rings of cartilage to prevent its collapse;
- dividing into two **bronchi** that repeatedly divide into **bronchioles**;
- the **alveoli** at the ends of the bronchioles where gaseous exchange occurs.

The **alveoli** are adapted as a gaseous exchange surface as:

- their shape and large number produce a **large surface area**;
- fluid lining the alveolus allows gases to dissolve and diffuse across;
- there is a short diffusion pathway:

  – a single layer of flattened epithelial cells – the alveolar wall, and

  – a single layer of endothelial cells – the capillary wall;

- an extensive network of blood capillaries surround each alveolus;
- ventilation and blood flow maintain a **high diffusion gradient**.

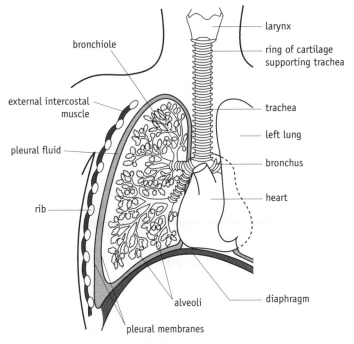

Structure of human lungs

An alveolus

✓ *Quick check 2*

Ventilation

Ventilation systems get gases to and from exchange surfaces.

Ventilation in mammals

Inspiration (breathing in) is an active process involving contraction of muscles.

Expiration (breathing out) is mainly a passive process when muscles relax. Elastic recoil of lung tissue raises the pressure above atmospheric.

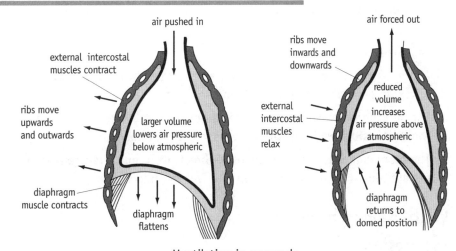

Ventilation in mammals

✓ *Quick check 3*

> Never say that air is sucked into the lungs. It is forced into the air passages – the trachea, bronchi and bronchioles – by higher external atmospheric pressure.

? *Quick check questions*

1 Describe the relationship between the size of organisms and the surface area : volume ratio.

2 Give two ways in which a respiratory surface is adapted for gaseous exchange.

3 Describe how inhalation takes place in mammals.

Gaseous exchange in other organisms

You need to know about gaseous exchange in fish and plants.

Gaseous exchange in fish

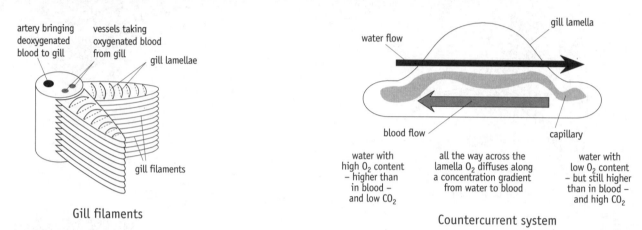

Gill filaments

Countercurrent system

In bony fish gas exchange occurs over the surface of the gills

- Each gill has two rows of **gill filaments** whose **lamellae** provide a **large surface area**.

- Lamellae have a lot of blood capillaries.

- A thin barrier of two cell layers (epithelial and endothelial) provides a short diffusion pathway between the blood and water.

- The ventilation mechanism provides a flow of water over the gills, maintaining a **high diffusion gradient** for oxygen and carbon dioxide.

- The blood and water flow in opposite directions (**countercurrent system**), maintaining a high concentration or diffusion gradient.

> Air goes in and out of our lungs but water flows in one direction over the gills. This is because water is too dense to easily breathe in and out – and contains less oxygen than air, so more has to pass over the exchange surface.

> ✓ *Quick check 1, 2*

Ventilation in bony fish

Inspiration

- The mouth opens and the floor of the **buccal cavity** (mouth) is lowered, increasing its volume.

- This decreases the pressure in the buccal cavity and water enters the mouth.

- At this time the **opercular valve** is shut and the **opercular cavity** is enlarged, reducing the pressure inside.

- Water is forced into the opercular cavity and over the gills from the buccal cavity.

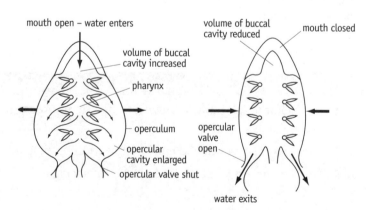

Ventilation in bony fish

Expiration

- The mouth is closed and the floor of the buccal cavity is raised.
- This reduces volume and increases pressure in the buccal cavity, forcing water over the gills, through the **gill slits** and into the opercular cavity.
- The pressure in the opercular cavity forces water out of the opercular valve.

✓ *Quick check 3*

Gaseous exchange in plants

A waterproof waxy cuticle covers leaves and allows very little gaseous exchange. Gaseous exchange involves stomata, mesophyll cells and the air spaces between them.

- **Stomata** are pores in the epidermis of leaves, surrounded by two guard cells.
- Gases diffuse in and out of leaves through the stomata.
- Numerous stomata with small diameter increase the rate of diffusion.
- Leaves are thin, producing a short diffusion pathway for gaseous exchange.
- Respiration and photosynthesis maintain diffusion gradients by using and producing oxygen and carbon dioxide.

> ▶ Only respiration takes place in leaf cells at night – so oxygen enters the leaf and carbon dioxide exits. In daylight, the cells still respire but they also carry out photosynthesis. When the rate of photosynthesis is greater than the rate of respiration, there is a net uptake of carbon dioxide and loss of oxygen.

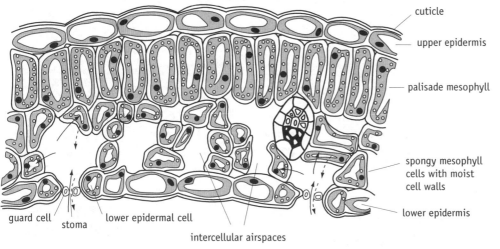

cuticle

upper epidermis

palisade mesophyll

⟶ gaseous
◀----- exchange

spongy mesophyll cells with moist cell walls

lower epidermis

guard cell   stoma   lower epidermal cell

intercellular airspaces

Gaseous exchange in leaves

- Numerous **mesophyll cells** lining the intercellular air spaces in the leaf provide a large surface area for gaseous exchange.
- Gases diffuse rapidly through the **intercellular air spaces** and dissolve in the moist cell walls of the mesophyll cells.
- Gases diffuse rapidly across the thin cell wall and cell membrane of mesophyll cells.

✓ *Quick check 4*

❓ Quick check questions

1 Give two features of the gill filaments that aid gaseous exchange.
2 Explain how the countercurrent system aids gaseous exchange in fish.
3 Describe how expiration occurs in a bony fish.
4 Describe how a carbon dioxide molecule in the atmosphere reaches the cytoplasm of a leaf mesophyll cell.

Enzymes

Enzymes are **globular proteins**. As **biological catalysts** enzymes:

- allow biochemical reactions to happen at the temperature found in an organism;
- are unaffected by reactions and can be used many times;
- have a specific tertiary structure and react with a specific **substrate** to produce a specific **product**.

Vital chemical reactions in living organisms need specific enzymes to proceed, so specific enzymes regulate biological processes.

> The key to understanding everything about enzymes is to remember they are proteins – with a very specific 3-dimensional shape. This allows them to 'recognise' other substances which have a shape that fits specifically against some part of the enzyme.

Lowering the activation energy

Lowering activation energy – energy needed to start a reaction – means that:

- chemical reactions in cells occur within an **acceptable temperature range**;
- the overall **rate of reaction** is **increased**.

Enzyme specificity

Enzymes are **highly specific.** Some act on a single substrate, others on particular chemical bonds.

- Enzyme specificity is due to the **tertiary structure** of an enzyme.
- This determines the shape (configuration) of the **active site**.

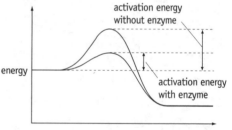

Lowering the activation energy

Lock and key hypothesis

- The **active site** is where the substrate binds.
- A substrate with a complementary shape binds and forms an **enzyme–substrate complex.**
- The reaction takes place and the product is released.
- The enzyme remains **unchanged**.

✓ Quick check 1

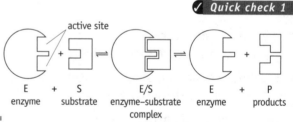

Lock and key hypothesis

Effect of temperature

- An increase in temperature results in **more collisions** between reactant molecules.
- The rate of reaction increases up to a maximum at the **optimum temperature**.
- Above the optimum, hydrogen and ionic bonds break – the enzyme is **denatured** – its tertiary structure has changed.
- The rate of reaction decreases as the substrate cannot attach to the active site.
- Denaturation of proteins at temperatures above 50°C is usually permanent.

> Read questions carefully – not all enzymes have an optimum temperature of 37°C and optimum pH of 7.

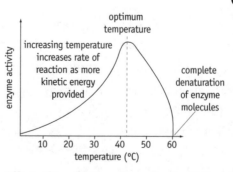

Effect of temperature on rate of reaction

Effect of pH

Enzymes possess an **optimum pH** at which the rate of reaction is at a maximum.
- Most enzymes work in a narrow pH range.
- A change in pH alters ionic charges of acidic and basic groups.
- The tertiary structure and active site are altered, so substrate cannot bind.
- Extremes of pH can cause **denaturation** (due to e.g. acid hydrolysis of the enzyme).

> ◗ Make sure you know that pH 7 is neutral, pH 1 is very acidic and pH 14 is very alkaline.

Effect of substrate concentration

Increasing the substrate concentration will increase the reaction rate to a point.
- The rate of reaction increases as collisions between substrate and enzyme molecules are more likely.
- The rate then levels out as active sites of all the enzyme molecules are taken up by substrate molecules.
- The rate is now limited by the time required for the enzyme/substrate complex to form and release the product.
- Adding more enzyme will increase the rate.

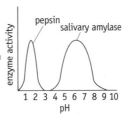

Effect of pH on rate of reaction

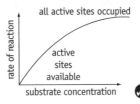

✓ *Quick check 2, 3*

Effect of substrate concentration on rate of reaction for a given concentration of enzyme

Enzyme inhibitors

Enzyme inhibitors slow down the rate of reaction.

Competitive inhibitor

- Has a **similar shape** to the substrate and **competes** for the **active site**.
- The rate of reaction is reduced, because of inhibitor occupying the active site.
- This inhibition can be overcome by adding more substrate – increasing the chance of substrate occupying the active site.

Non-competitive inhibitor

- Binds at a site other than the active site, **altering the shape** of the enzyme and its **active site**.
- The substrate cannot attach, or the substrate binds but no product is formed.
- The addition of more substrate will not reduce this inhibition.

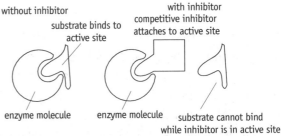

Competitive inhibition

Non-competitive inhibition

✓ *Quick check 4*

? *Quick check questions*

1 Explain why lipases can hydrolyse lipids but not carbohydrates.

2 Give three factors affecting the rate of an enzyme-catalysed reaction.

3 Explain in terms of molecular shape the effect of high temperature on enzyme activity.

4 Suggest a simple method to determine whether a reaction is being inhibited by a competitive or non-competitive inhibitor.

Digestion

Digestion is the hydrolysis of large, insoluble organic compounds into small, soluble molecules that can be absorbed into cells. This involves **hydrolytic** enzymes.

✓ *Quick check 1*

Extracellular digestion – saprophytic fungi

Saprophytic fungi are **decomposers**, feeding and growing on dead organic matter. This is essential for the recycling of nutrients in an ecosystem.

- Most fungi consist of a mass of branching threads called **hyphae.**
- The whole mass of hyphae is known as the **mycelium.**
- Enzymes are **secreted** by the hyphae for **extracellular** digestion.
- These enzymes include **carbohrases**, **lipases** and **proteases**.
- The enzymes secreted determine what the fungi can feed on.
- Large, insoluble organic compounds are hydrolysed/digested into small, soluble molecules.
- These are **absorbed** through the cell membrane of the hyphae.
- The mycelium provides a large surface area of hyphae for absorption.

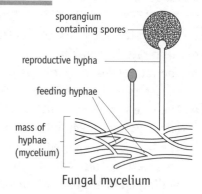

Fungal mycelium

▶ Look to mention: secretion of digestive enzymes, digestion/hydrolysis of food, absorption of soluble products.

Starch-agar plates

These can be used for assaying the **carbohydrase** activity of saprophytic fungi, specifically **amylase** which breaks down starch.

Method

- Starch is mixed with molten agar before pouring the mixture into Petri dishes.
- The **starch-agar** is left to cool and solidify.
- Equal sizes of **mycelial discs** from different species of fungi are then placed on separate starch-agar plates.
- The plates are kept at the same temperature for 24 hours.
- The plates are then flooded with **potassium iodide solution** and the diameter of the clear zone surrounding each fungus is measured.

Starch-agar plates

The diameter of the **clear zone** shows where no starch is present. It is produced because amylase is secreted from the hyphae and **hydrolyses** the surrounding starch into maltose.

✓ *Quick check 2*

The human gut

The generalised structure of the gut wall is similar throughout the alimentary canal, but is modified in different parts of the gut according to its function.

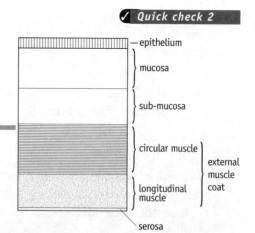

Generalised structure of the human gut wall

The oesophagus

The oesophagus connects the mouth to the stomach. After swallowing, the food (bolus) is pushed down the oesophagus by **peristalsis**:

- A wave of contraction of circular muscle behind the bolus and relaxation of longitudinal muscles;

- mucus secreted by glands in the oesophagus lining lubricates the passage of food.

▶ In the exam you may be asked to label the different layers of the gut wall.

✓ *Quick check 3*

▶ Peristalsis also moves food through the intestines

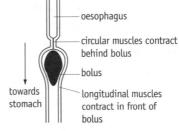

oesophagus

circular muscles contract behind bolus

bolus

towards stomach

longitudinal muscles contract in front of bolus

Peristalsis

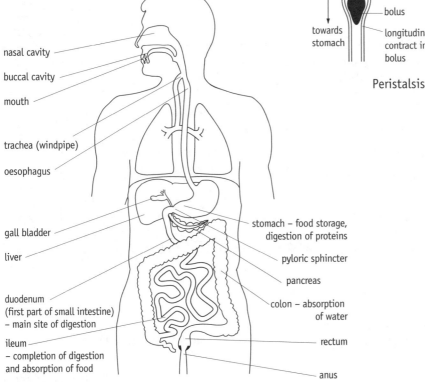

nasal cavity

buccal cavity

mouth

trachea (windpipe)

oesophagus

gall bladder

liver

duodenum (first part of small intestine) – main site of digestion

ileum – completion of digestion and absorption of food

stomach – food storage, digestion of proteins

pyloric sphincter

pancreas

colon – absorption of water

rectum

anus

Human digestive system

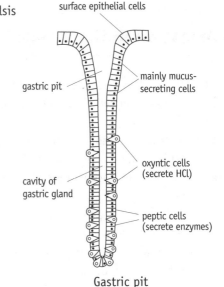

surface epithelial cells

gastric pit

mainly mucus-secreting cells

oxyntic cells (secrete HCl)

cavity of gastric gland

peptic cells (secrete enzymes)

Gastric pit

The stomach

The muscular layers of the stomach mix the food with the **gastric juice** which is secreted by gastric pits in the stomach mucosa.

The intestines

The different parts of the intestines are adapted for the secretion of digestive juices and the absorption of digested food molecules.

Digestion occurs in the duodenum and ileum. Most absorption occurs in the ileum.

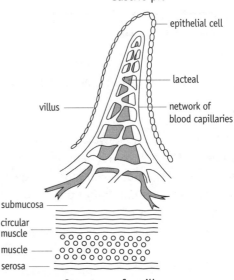

epithelial cell

lacteal

network of blood capillaries

villus

submucosa

circular muscle

muscle

serosa

Structure of a villus

? Quick check questions

1 Explain why the digestion of food molecules is necessary.

2 Briefly describe how you would compare the amylase activity of two species of fungi.

3 Describe how peristalsis takes place.

Digestion in humans

Digestion in humans involves physical (e.g. chewing) and chemical (enzymic) processes. Enzymes are secreted into the gut to hydrolyse the food molecules.

Digestion in the mouth

The sight, smell and taste of food stimulate secretion of saliva from salivary glands.
- Saliva contains water, the enzyme **salivary amylase** and mucus.
- Salivary amylase **hydrolyses** starch into maltose.

> ▶ Chewing (mastication) breaks the food into smaller particles, increasing the surface area for enzymes to act on so that digestion is faster.

Digestion in the stomach

The stomach lining produces **gastric juice** containing the enzyme **pepsin, hydrochloric acid** (HCl) and **mucus.**
- **Pepsin** is an **endopeptidase** – it **hydrolyses** peptide bonds in the middle of **proteins**.
- Pepsin has an optimum pH of around 1.8 which is produced by HCl.
- Food in the stomach is churned by peristalsis into an acidic fluid, **chyme.**
- At intervals, small amounts of chyme are released into the duodenum through the **pyloric sphincter.**

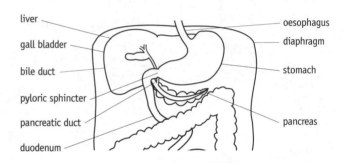

liver — oesophagus
gall bladder — diaphragm
bile duct — stomach
pyloric sphincter
pancreatic duct — pancreas
duodenum

Pancreas and gall bladder

Digestion in the duodenum

The duodenum is the beginning of the small intestine. Alkaline bile and pancreatic juices neutralise acidic chyme from the stomach.

> ▶ Mucus produced in the stomach and throughout the alimentary canal aids the passage of food and helps to prevent breakdown of the mucosa

Bile

Bile contains sodium hydrogen carbonate and **bile salts.** It is produced by the liver, stored in the gall bladder and enters the duodenum by the bile duct.
- It provides an optimum pH for digestive enzymes.
- The bile salts **emulsify** lipids into small fat droplets, increasing the surface area for the action of lipase.
- This speeds up hydrolysis of lipids by **pancreatic lipase.**

Pancreatic juice

This contains sodium hydrogen carbonate and enzymes including:
- **amylase** – hydrolysis of starch into maltose;
- **lipase** – hydrolyses lipids into fatty acids and glycerol;
- **trypsin (endopeptidase)** – hydrolyses polypeptides into smaller polypeptides;
- **exopeptidases** – hydrolyse peptide bonds at the ends of polypeptides, removing amino acids and producing dipeptides.

endopeptidases hydrolyse peptide bonds in the middle of a polypeptide chain, producing smaller polypeptides

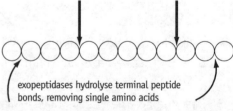

exopeptidases hydrolyse terminal peptide bonds, removing single amino acids

Action of endopeptidases and exopeptidases

Digestion in the duodenum and ileum

The mucosa produces a slightly alkaline fluid (intestinal juice) that contains mucus, water, sodium hydrogen carbonate and several enzymes including:

- **exopeptidases** – continue to hydrolyse polypeptide chains;
- **dipeptidases** – hydrolyse dipeptides to produce amino acids;
- **lipases** – hydrolyse lipids to fatty acids and glycerol;
- **maltase** – hydrolyses maltose (disaccharide) to two glucose molecules.

Absorption in the ileum

Absorption of digested food products occurs in the duodenum and to a greater extent in the ileum. The ileum is adapted for absorption in the following ways:

- **large surface area** due to its long length, presence of **villi** and **microvilli**;
- villi lined by a **single layer** of epithelial cells – **short diffusion pathway**;
- **moist lining** enabling soluble substances to dissolve and pass through;
- villi contain **blood capillaries** which carry away digested products – maintaining a **high diffusion gradient** for absorption;
- **lacteals** in the villi absorb digested lipids maintaining a **high diffusion gradient**;
- **epithelial cells** have many **mitochondria** – supply **ATP** for active transport;
- **carrier proteins** are present in membranes, increasing permeability.

Absorption is by **diffusion**, **facilitated diffusion** and **active transport**.

Mechanism for absorption		
Diffusion	**Facilitated diffusion**	**Active transport**
Water	Amino acids	Mineral ions
Fatty acids	Glucose	Amino acids
Glycerol		Glucose

The enzymes that break down disaccharides (e.g. maltase) and dipeptidases are part of the cell surface membrane of epithelial cells.

> ▶ Digestive enzymes work together to speed up digestion. Endopeptidases produce many shorter polypeptides with lots of 'ends' for exopeptidases to work on – this produces many dipeptides for dipeptidases to work on. Can you see a similar pattern for digestion of starch?

> ✓ *Quick check 1,2,3*

> ▶ Absorption involves movement of substances across cell membranes – see the earlier section about mechanisms involved.

> ✓ *Quick check 4*

> **?** ## Quick check questions
>
> 1 Name the two enzymes involved in the complete hydrolysis of starch to glucose.
>
> 2 Explain the role of bile in the digestion of lipids.
>
> 3 Produce a flow chart to show the action of enzymes in the hydrolysis of a protein into individual amino acids.
>
> 4 Describe and explain how the ileum is adapted for the absorption of digested food products.

Module 1: end-of-module questions

1 a Explain how the structure of each of the following carbohydrates is related to its function: (i) cellulose; (ii) starch. [4]

b Use diagrams to describe how fatty acids and glycerol combine to form a triglyceride. [3]

c Explain how the primary structure of a polypeptide leads to its tertiary structure. [4]

2 a Describe how you would test a solution for the presence of each of the following substances: (i) sucrose; (ii) protein; (iii) lipid. [3]

b You are given a solution containing a soluble protein. Suggest how you would find out which amino acids were present in this protein. [4]

3 a i Describe the differences you would expect to see between an epithelial cell from the small intestine and a palisade mesophyll cell from a plant when viewed with a light microscope. [3]

ii Name **three** organelles that you would expect to see in both types of cell with an electron microscope, and describe the function of each. [6]

b When scientists first saw mitochondria with an electron microscope they had no idea what their function was. Explain how they were able to find the function of mitochondria in cells that contain many different types of organelle. [3]

4 a The diagram shows three plant cells and their water potentials.

i Use arrows to show the directions of net water movement between these cells. [1]

ii Explain why net movements of water would take place. [3]

b The chart shows concentrations of ions in the sap of a plant cell and in the solution surrounding the cell.

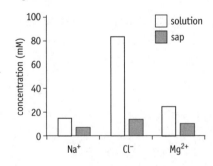

    i Explain the concentration of chloride ions in the sap compared with the external solution. [2]

    ii Suggest why the amount of accumulation of each of the ions is different. [2]

    iii The cell was placed into a new solution with the same concentrations of ions, but half the concentration of dissolved oxygen. Suggest the effect that this would have on the results shown above. [3]

5 a Explain what is meant by each of the following:

    i diffusion; [2]

    ii facilitated diffusion; [3]

    iii active transport. [3]

b Explain why water can cross the cell membrane by diffusion but glucose needs a carrier protein. [5]

6 a The diagram shows cross-sections of the bodies of two species of animal.

The animals have the same body lengths and live in the same environment. Animal **X** exchanges gases across its whole body surface. Animal **Y** has specialised internal gas exchange surfaces. Suggest an explanation for the differences in gas exchange surfaces between X and Y. [3]

b Explain how the leaf of a dicotyledonous plant is adapted for the exchange of gases. [4]

c Describe the similarities between the gas exchange surfaces in mammals and fish. [4]

7 a Explain the importance of enzymes to the chemical reactions necessary to support life. [3]

b The graph shows the effects of increasing substrate concentration on the rate of action of an enzyme with and without an inhibitor.

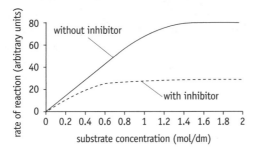

    i Explain the results without the inhibitor. [3]

    ii Describe and explain the effect of the inhibitor. [3]

8 The graph shows the effect of temperature on the activity of an enzyme.

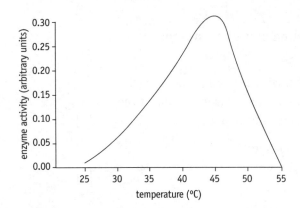

a Give the optimum temperature for this enzyme. [1]

b i Explain the activity of the enzyme at 25°C. [2]

ii Explain the activity of the enzyme at 55°C. [3]

c Two samples of the enzyme were taken, both were stored at 5°C for several days but one was boiled before being stored. Both samples were then warmed to the optimum temperature for the enzyme. Substrate was added and the activity of the enzyme in each sample was measured.

i Explain any differences in the results you would expect with each sample. [2]

ii Explain three factors you would need to keep constant to make accurate comparisons about the effects of temperature on the samples. [6]

9 a Explain how protein is digested in the gut. [6]

b Describe how the structure of the wall of each of the following is adapted to its function: (i) the oesophagus; (ii) the stomach; (iii) the small intestine. [12]

c Sometimes a person has to have their gall bladder removed. This means that bile is not stored and released when required after a meal, but is released in a constant, slow trickle into the small intestine. Suggest how this operation would affect digestion of a fat-rich meal. [3]

10 a Explain how starch is digested in the gut. [6]

b Describe how the products of starch digestion are absorbed in the small intestine. [5]

Module 2: Genes and Genetic Engineering

This module is broken down into four topics: The genetic code; The cell cycle; Sexual reproduction; and Applications of gene technology.

The genetic code

- Information on how to make and run cells and organisms is passed from one generation to the next by DNA, as a sequence of bases and using the genetic code.
- In eukaryotic cells, each DNA molecule is part of a chromosome in the nucleus.
- The structure of DNA allows it to carry coded genetic information, and to replicate itself and the information.
- Specific base pairing between the bases (nucleotides) that make up the double-helix of the molecule is necessary for replication of DNA and use of the information by the cell.
- A gene is a length of DNA carrying coded information for making a specific protein (often an enzyme).
- A gene can exist in different forms known as alleles, each producing a different form of the protein. New alleles are formed by mutation.
- Making a protein involves taking a copy of the information in DNA (transcription) and using this to make protein at the ribosomes (translation).

The cell cycle

- New cells come from existing cells by cell division.
- The growth of organisms and asexual reproduction depend on mitosis. This produces new cells which are genetically identical to the parent cell.
- Clones are genetically identical organisms produced by processes involving mitosis.

Sexual reproduction

- Most organisms have some form of sexual reproduction, involving the fusion of gametes at fertilisation.
- Cell division by meiosis halves the number of chromosomes in the cells produced. The fusion of gametes reverses this process and restores the usual chromosome number. If meiosis did not take place, the number of chromosomes in cells would double with every generation.

Applications of gene technology

- Genetic engineering techniques allow us to remove genes from the DNA of one organism and insert them into the DNA of another, even if it is a different species.
- We can make copies of DNA using the polymerase chain reaction. Organisms which receive a gene are genetically engineered, or modified.
- Microorganisms are often genetically engineered and cultured on an industrial scale.
- Attempts are also being made to insert genes into humans to treat genetic disorders caused by defective genes, e.g. cystic fibrosis. Some animals and plants have been genetically modified to make things useful to humans. This technology raises ethical problems.

Structure of nucleic acids

DNA (deoxyribonucleic acid) and RNA (ribonucleic acid) are nucleic acids. They are **polymers** of **nucleotides**. DNA consists of two **polynucleotide** strands, whereas RNA consists of a single polynucleotide strand. Each nucleotide consists of three molecules joined by **condensation** reactions:

- a **five-carbon sugar** (pentose);
- a **phosphoric acid** molecule;
- a **nitrogen-containing organic base**.

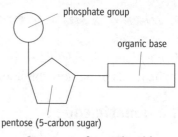

Structure of a nucleotide

DNA

Structure of DNA

The two polynucleotide strands are held together by **hydrogen bonding** – forming a **double helix**. In DNA, four bases are found in the nucleotides: **cytosine** and **thymine** (pyrimidines); **adenine** and **guanine** (purines).

- The sugar **deoxyribose** and **phosphate** form the backbone of the polynucleotide strands.

- The bases are orientated towards the centre of the helix, protecting them from reacting with other chemicals.

- Bases on one strand have **specific base pairing** with bases on the other strand.

> **Adenine always pairs with Thymine**
> **Guanine always pairs with Cytosine**

- Bases are joined by weak **hydrogen bonds**, but there are so many between the strands that together they make DNA a **stable polynucleotide**.

- The DNA helix is further coiled to produce a super helix, providing a compact store of genetic information.

> In almost any question concerning DNA or RNA, specific base pairing needs to be referred to – and understood. You may have been taught this as complementary base pairing.

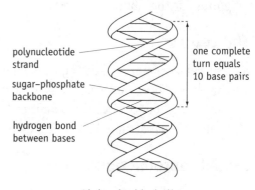

Alpha double helix

✓ *Quick check 1,2,3*

The role of DNA

- It carries the **genetic code** controlling the synthesis of proteins. By controlling which proteins, particularly enzymes, are produced in a cell, DNA controls the development, structure and function of a cell.

- It is capable of **self-replication**, which is essential for increases in cell number during growth, reproduction (asexual and sexual), and passing genetic information to the next generation.

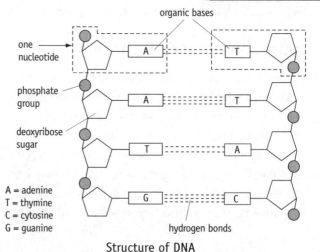

A = adenine
T = thymine
C = cytosine
G = guanine

Structure of DNA

- Although DNA is a relatively stable molecule, alterations in the genetic information (**mutations**) can occur, providing **variation** – the basis for evolution via natural selection.

> ▶ You will need to know similarities and differences in structure between DNA and RNA.

RNA

Structure of RNA

The structure of RNA differs from DNA in that:

- the pentose is **ribose** not deoxyribose;
- the bases found in RNA are adenine, guanine, cytosine and **uracil**; uracil replaces thymine and binds to adenine.
- it is **single-stranded**, not double-stranded;
- it is **shorter** than DNA, with a lower molecular mass.

Types of RNA

There are three types of RNA. Ribosomal RNA is a structural component of ribosomes. Messenger RNA (mRNA) and transfer RNA (tRNA) have important roles in protein synthesis.

> ▶ DNA is much too large to be soluble, or to pass through the pores in the nuclear envelope – it stays in the nucleus. mRNA and tRNA are small enough to be soluble and move around the cytoplasm of the cell.

Messenger RNA (mRNA)

- This is a single polynucleotide strand formed in the nucleus during **transcription**, using a specific section of a DNA molecule (a **gene**) as a blueprint or template.
- mRNA carries a 'copy' of the genetic information of the gene to ribosomes in the cytoplasm.
- The mRNA is used in **translation** to determine the sequence of amino acids in a protein (its primary structure).

Transfer RNA (tRNA)

- A tRNA molecule is a single strand, folded into a 'clover leaf' shape.
- There are different types of tRNA molecule in the cytoplasm, each with a binding site for the attachment of a **specific amino acid**.
- During protein synthesis each tRNA molecule carries its specific amino acid to a ribosome.
- A specific sequence of three bases on the molecule is known as the **anticodon**.

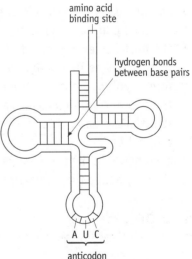

amino acid binding site

hydrogen bonds between base pairs

A U C

anticodon
messenger RNA binding site

Structure of tRNA

> ✓ *Quick check 4*

? ## Quick check questions

1 Give the components of a nucleotide.

2 Name the four organic bases present in DNA.

3 How are the bases joined together in DNA?

4 Give two ways in which the structure of: (i) RNA differs from DNA; (ii) tRNA differs from mRNA.

DNA – replication and the genetic code

DNA replication occurs as part of the process of cell division, and is essential for the growth and reproduction of organisms.

The semi-conservative mechanism of DNA replication

When DNA replicates:
- its double helix uncoils (unzips) into two separate strands as hydrogen bonds between the polynucleotide strands are broken;
- each strand acts as a **template** for the formation of a new complementary strand;
- **nucleotides** bind to each template strand by **specific base pairing**;
- adenine pairs with thymine and cytosine pairs with guanine;
- the nucleotides are joined together by the enzyme **DNA polymerase** to form a **polynucleotide strand**;
- the two new DNA molecules are **identical** to each other and to the original DNA;
- each newly formed DNA molecule contains one of the original polynucleotide strands, hence the term **semi-conservative replication**.

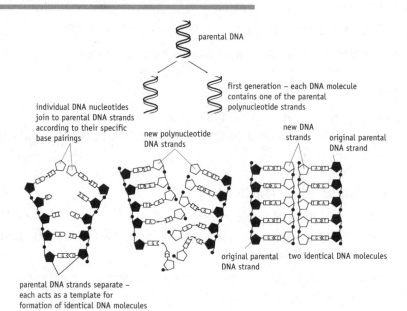

Semi-conservative mechanism of DNA replication

✓ *Quick check 1, 2*

The gene

Genes are lengths of DNA which contain coded genetic information that determines the nature and development of organisms.
- A gene is the **sequence of (nucleotide) bases of DNA** coding for the production of a **specific polypeptide** by determining the sequence of amino acids (primary structure).
- Genes are located along **chromosomes** – thread-like structures consisting of DNA and protein.
- A gene can exist in different forms, called **alleles**, that code for different types of the same characteristic.
- Alleles of a particular gene are located in the same relative position (**locus**) on homologous chromosomes.
- A homologous pair of chromosomes carry the same genes, but not necessarily the same alleles of those genes.

> Gene and allele do not mean the same thing – you need to be careful which you refer to.

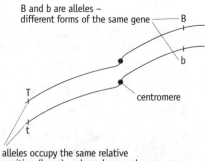

alleles occupy the same relative position (locus) on homologous chromosomes

A pair of homologous chromosomes

✓ *Quick check 3, 4*

The genetic code

DNA carries **genetic information** that determines the sequence of amino acids in proteins. The **genetic code** is founded on sequences of (**nucleotide**) **bases**.

- A sequence of three (nucleotide) bases is called a **base triplet**.
- The presence of four different nucleotides in DNA and RNA means there are 64 (4^3) possible base triplets.
- These code for the 20 commonly occurring amino acids in living organisms.
- The base triplets of mRNA are known as **codons**.
- The genetic code is **degenerate** as some amino acids are coded for by more than one codon, e.g. six codons code for the amino acid arginine.

Some codons are used for 'punctuation'; for example, a 'stop' or nonsense codon codes for the end of a particular gene.

> The genetic code is universal, with all organisms using the same nucleotide bases to code for the same amino acids.

✓ *Quick check 5*

		second base							
		U		C		A		G	
U	UUU	Phe	UCU	Ser	UAU	Tyr	UGU	Cys	U
	UUC	Phe	UCC	Ser	UAC	Tyr	UGC	Cys	C
	UUA	Leu	UCA	Ser	UAA	Nonsense	UGA	Nonsense	A
	UUG	Leu	UCG	Ser	UAG	Nonsense	UGG	Try	G
C	CUU	Leu	CCU	Pro	CAU	His	CGU	Arg	U
	CUC	Leu	CCC	Pro	CAC	His	CGC	Arg	C
	CUA	Leu	CCA	Pro	CAA	Gln	CGA	Arg	A
	CUG	Leu	CCG	Pro	CAG	Gln	CGG	Arg	G
A	AUU	Ile	ACU	Thr	AAU	Asn	AGU	Ser	U
	AUC	Ile	ACC	Thr	AAC	Asn	AGC	Ser	C
	AUA	Ile	ACA	Thr	AAA	Lys	AGA	Arg	A
	AUG	Met	ACG	Thr	AAG	Lys	AGG	Arg	G
G	GUU	Val	GCU	Ala	GAU	Asp	GGU	Gly	U
	GUC	Val	GCC	Ala	GAC	Asp	GGC	Gly	C
	GUA	Val	GCA	Ala	GAA	Glu	GGA	Gly	A
	GUG	Val	GCG	Ala	GAG	Glu	GGG	Gly	G

(first base on left, third base on right)

> You don't need to learn these

Quick check questions

1 Explain what is meant by semi-conservative replication.
2 What is the role of the enzyme DNA polymerase during DNA replication?
3 What is an allele?
4 Give a precise definition of a gene.
5 Explain what is meant by a degenerate genetic code.

DNA and protein synthesis

Protein synthesis can be divided into two processes: **transcription** and **translation**. Transcription occurs in the **nucleus** and involves 'rewriting' (transcribing) part of the DNA code into a strand of messenger RNA. Translation occurs in the cytoplasm and involves ribosomes synthesising proteins using the information provided by messenger RNA.

Transcription

During transcription:

- the section of the DNA molecule (a gene) uncoils and the two polynucleotide strands separate as hydrogen bonds are broken;

- one strand is the template or **sense strand**;

- RNA nucleotides line up alongside the DNA nucleotide bases on the template strand by specific (complementary) base pairing, with uracil (RNA) pairing with adenine (DNA);

- the enzyme **RNA polymerase** joins the nucleotides together to form a strand of messenger RNA.

The mRNA strand leaves the nucleus through a nuclear pore and attaches to a ribosome in the cytoplasm, where translation occurs. The strands of DNA in the nucleus will recoil when sufficient mRNA has been produced.

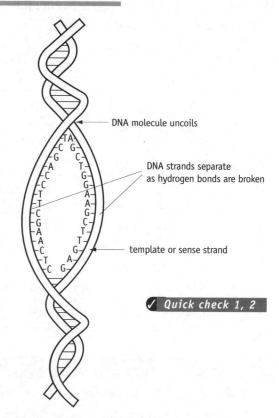

DNA molecule uncoils

DNA strands separate as hydrogen bonds are broken

template or sense strand

✓ *Quick check 1, 2*

Transcription – DNA strands separate

Translation

During translation, the sequence of codons on the mRNA strand determines the sequence of amino acids in a polypeptide. Translation is carried out at ribosomes in the cytoplasm. In the cytoplasm there is a specific type of tRNA for each of the twenty amino acids. Each tRNA molecule has three exposed bases known as an **anticodon**.

- An mRNA attaches to a ribosome.

- A tRNA molecule with the complementary anticodon binds to the first codon on the mRNA strand, bringing its specific amino acid.

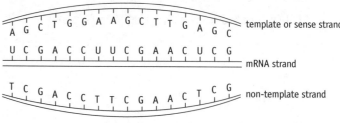

template or sense strand

mRNA strand

non-template strand

Transcription – formation of RNA

- The **anticodon** binds to the **codon** by **specific (complementary) base pairing**.
- Another tRNA then binds to the second codon on the mRNA strand.
- The amino acid on the first tRNA molecule is joined to the amino acid on the second tRNA molecule by a peptide bond.
- This requires ATP and the action of an enzyme.
- The first tRNA molecule then moves away from the ribosome leaving its amino acid behind, and the mRNA moves over the ribosome by one codon.

As the mRNA strand has been transcribed from the DNA template strand, it is the sequence of DNA (nucleotide) bases that ultimately determines which specific polypeptide is produced.

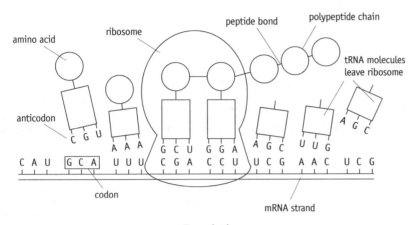

Translation

- This process continues along the mRNA strand until all the codons have been 'read' and the primary structure of the polypeptide has been produced.
- The polypeptide folds itself into its secondary and tertiary structure.
- The sequence of amino acids in this polypeptide has been determined by the sequence of codons on the mRNA strand.

✓ *Quick check 3, 4*

Quick check questions

1 Give the RNA sequence which would be complementary to the following DNA bases: T T G C G A C G T G G C A

2 Give two ways in which DNA replication and transcription differ.

3 Name the product of translation.

4 Describe the role of tRNA molecules during translation.

Gene mutations

Gene mutations are **changes in the sequence** of (nucleotide) bases in DNA, resulting in the formation of a different polypeptide. The altered base sequence codes for a different sequence of amino acids. **New alleles** of genes are produced by mutations.

Causes of mutations

- Mutations occur naturally **at random** and the rate varies from gene to gene.
- Mutations can arise as a result of incorrect pairing during DNA replication.
- Most mutations are harmful and recessive, although dominant and beneficial mutations can also arise.
- **Mutagenic agents** increase the frequency of mutations, e.g.
  - high energy (ionizing) radiation, e.g. X-rays, gamma rays, UV light
  - high energy particles, e.g. alpha and beta particles
  - chemicals such as nitrous oxide, benzene.

> ▶ It is not enough to say that 'radioactivity' or 'radiation' cause mutations – you need to give the specific examples here.

> ✓ **Quick check 1**

Types of gene mutations

There are several ways in which the sequence of nucleotide bases in a gene can be altered, including substitution, deletion and addition.

Substitution

Substitution is the **replacement** of one or more bases by one or more different bases. The substitution of a single base may result in:

- a new codon coding for a different amino acid in the polypeptide chain which may result in a **non-functional protein** being formed;
- the same amino acid being coded due to the degeneracy of the DNA code so that the **polypeptide remains unchanged**;
- the formation of a **nonsense codon** which terminates the polypeptide chain so that a non-functional protein is produced.

> ▶ Deletions and additions are usually more harmful than substitutions, as more amino acids are affected in the protein.

Deletion

Deletion is the **removal** of one or more bases resulting in:

- a **frame shift**, which is the alteration in the codons from the point of deletion;
- a different sequence of amino acids after the point of deletion and the formation of a different protein (often non-functional).

normal DNA   T A G   C C A   (T) A A   C G C   A G T

amino acid sequence   Ile — Gly — Ile — Ala — Ser

mutant DNA   T A G   C C A   A A C   G C A   G T

amino acid sequence   Ile — Gly — Leu — Arg — His

'frame shift'
amino acid sequence changed from point of deletion

Example of a deletion

Addition

This is the **addition** of one or more bases, resulting in:

- a **frame shift** – a change in the codons from the point of addition;
- a different sequence of amino acids after the point of addition and the formation of a different protein (often non-functional).

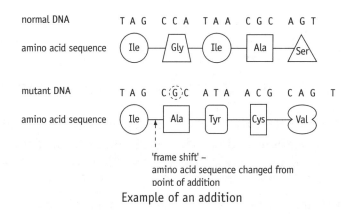

'frame shift' – amino acid sequence changed from point of addition

Example of an addition

✓ **Quick check 2, 3**

Effects of mutations

Many of the polypeptides/proteins coded for by genes are **enzymes**, and these regulate biological processes. A gene mutation that alters the sequence of amino acids in a polypeptide alters the primary structure of an enzyme and its function. The amino acid(s) changed may be:

- at the active site, changing its shape so that the substrate **cannot attach** – making the enzyme **non-functional**;
- at the active site, but the substrate can still attach although less effectively – **slowing down** enzyme activity;
- away from the active site, but affects the whole **tertiary structure** and shape of the enzyme and its active site and thus still affects enzyme activity.

Metabolic pathways involve many enzymes, and a mutation affecting the activity of one of these enzymes can block a whole metabolic pathway.

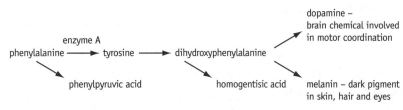

absence of enzyme A can result in lack of pigmentation and poor motor coordination

Example of a metabolic block

? ### Quick check questions

1 Give two examples of mutagenic agents.
2 Explain how a gene mutation involving the substitution of a base may have no effect on the polypeptide coded for.
3 Explain why a gene mutation involving the deletion of a base usually results in a non-functional polypeptide being produced.

Mitosis

Mitosis is the type of cell division that produces genetically identical cells. During mitosis DNA replicates in the parent cell, which divides to produce two new cells, each containing an exact copy of the DNA in the parent cell. The only source of variation in the cells is via mutations.

- Mitosis increases cell numbers during growth, repair of tissues and asexual reproduction.

- During mitosis, the nuclear material becomes visible as structures called **chromosomes**.

- In a normal body cell (somatic cell) the chromosomes can be grouped into **homologous pairs** of chromosomes.

- **Diploid number (2n)** is the total number of chromosomes in a normal body cell. In man this is 46, i.e. 23 homologous pairs.

- **Haploid number (n)** is a **single set of chromosomes**, i.e. one member from each homologous pair. In man this is 23, the number of chromosomes in a **gamete** (sperm or ovum).

- Mitosis produces cells with the same number of chromosomes as the parent cell so that a diploid parent cell will divide to produce two identical diploid cells.

> The diploid number of chromosomes is a characteristic of a particular species – and it isn't usually 46!

> Homologous chromosomes carry the same genes but may not carry the same alleles of the genes.

> ✓ *Quick check 1, 2*

Process of cell division

Biologists have divided the process of cell division into a number of stages: interphase, prophase, metaphase, anaphase and telophase.

> In rapidly dividing cells mitosis is completed within 24 hours.

Interphase

- This is when the cell is **not dividing,** but is carrying out its normal cellular functions.

- During interphase the cell prepares for nuclear division by:

  – **DNA replication**, doubling the genetic content of the cell

  – **ATP** content is increased, as nuclear division is a very active process.

- Replication of cell organelles, e.g. mitochondria, occurs in the cytoplasm.

Prophase

- The previously indistinct nuclear material is now visible as **chromosomes**.

- Due to DNA replication during interphase, each chromosome consists of two identical **sister chromatids** connected at the **centromere**.

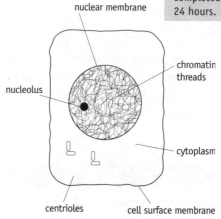

Interphase

Prophase

- Each chromosome shortens and thickens, a process known as **condensation**.
- Condensation of chromosomes prevents tangling with other chromosomes.
- Centrioles (in animal cells) move to opposite poles (sides) of the cell.
- The nucleoli and **nuclear membrane** break down.

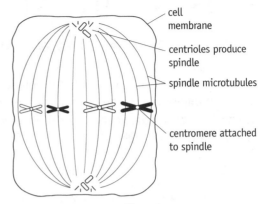

cell membrane

centrioles produce spindle

spindle microtubules

centromere attached to spindle

Metaphase

Metaphase

- A **spindle** (of protein microtubules) forms across the cell.
- Each chromosome moves to the equator of the spindle and attaches to a spindle fibre by its **centromere**.
- The **sister chromatids** of each chromosome are orientated towards opposite poles of the cell.

Anaphase

- The centromere splits and the **sister chromatids separate**.
- Sister chromatids move to opposite poles of the **spindle**.
- Numerous mitochondria around the spindle provide energy for movement.

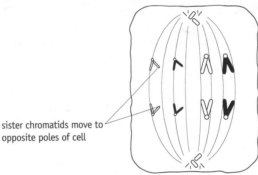

sister chromatids move to opposite poles of cell

Anaphase

> ◖ Once the chromatids reach the poles of the spindle they are called chromosomes. They will replicate themselves during the next interphase to produce two sister chromatids – ready for the next cell division.

Telophase

- The chromatids are at opposite poles of the cell.
- A nuclear membrane forms around each set of chromatids.
- The two nuclei formed are **genetically identical** to each other and the original parent cell.
- Two new cells are formed as **cytoplasmic cleavage** occurs and a cell membrane forms between the cells.

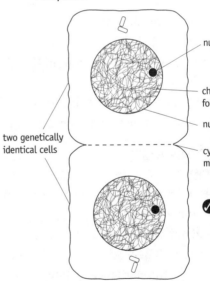

nucleolus forms

chromatids unwind to form chromatin

nuclear envelope forms

two genetically identical cells

cytoplasm divides and cell membrane forms

✓ *Quick check 3*

Telophase

? Quick check questions

1 Give two ways in which mitosis is important in living organisms.

2 Explain what the diploid number means.

3 During which stage of mitosis does each of the following occur:
   (i) separation of chromatids; (ii) condensation of chromosomes;
   (iii) DNA replication.

Meiosis and sexual reproduction

During meiosis in diploid organisms such as humans, cells containing pairs of homologous chromosomes divide to produce **haploid gametes**, containing one chromosome from each homologous pair. The cells produced are genetically different to the parent cell and to each other.

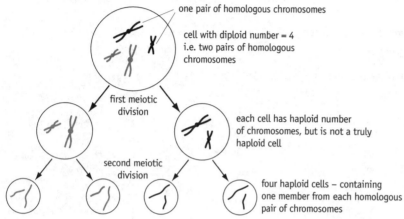

Process of meiosis

> Do not learn lots of detail about what happens in meiosis – it will not be asked about in the AS exam. These details are in the A2.

In meiosis:

- the DNA in a cell replicates only once;
- but the cell then divides twice;
- the number of chromosomes is reduced from the diploid number ($2n$) to the haploid number (n);
- four new cells are formed which are **genetically different** to each other;
- the cells produced usually function as **gametes**, i.e. reproductive cells.

Importance of meiosis

- In diploid organisms, meiosis is important in sexual reproduction as it ensures that **haploid gametes** are produced.
- These fuse at **fertilisation** to form a **zygote**, and the **diploid number** is restored.
- Meiosis ensures that each generation possesses a **constant number of chromosomes**, e.g. 46 chromosomes in man.
- If meiosis did not occur and diploid gametes were produced, the number of chromosomes would double every generation after fertilisation.
- The process of meiosis produces genetic variation in gametes.

Gametes

Sexual reproduction involves gamete formation and fertilisation. In sexual reproduction, DNA from one generation is passed to the next generation by gametes. Gametes are sexual reproductive cells and are haploid.

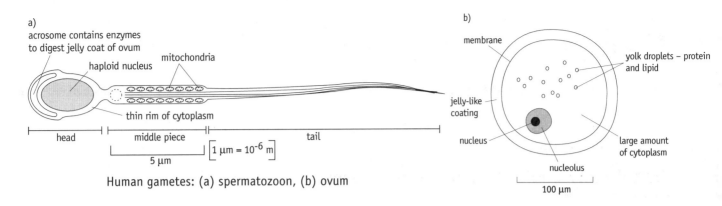

Human gametes: (a) spermatozoon, (b) ovum

- Although similar in some organisms, most species have clearly different male and female gametes.

- Female gametes are usually larger, with more cytoplasm and food reserves than male gametes.

- In many species male gametes are usually produced in considerably larger numbers than female gametes.

- Male gametes are mobile, often swimming to the female gamete using a 'tail'.

> The human sperm contain a large number of mitochondria to provide the energy required for swimming to the ovum.

✓ *Quick check 1, 2*

Life cycles

In a life cycle involving sexual reproduction there must be a haploid and a diploid stage. However, the timing of meiosis varies in different species to produce different life cycles. Two examples of different life cycles are shown on the right. Do not assume that meiosis produces the gametes. Look for where the chromosome number halves from diploid to haploid – that's when meiosis takes place.

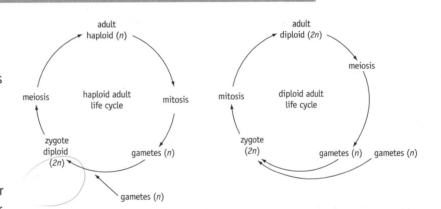

Types of life cycles

✓ *Quick check 3*

1 Give two differences between male and female gametes in a human.

2 What is a gamete?

3 Explain how meiosis enables a constant number of chromosomes to be maintained from generation to generation.

Genetic engineering

In genetic engineering, genes ('foreign DNA') are taken from one organism and inserted into another.

Genetically engineered microorganisms

Microorganisms, particularly bacteria, are widely used as **recipient** cells during **gene transfer**. The rapid reproduction of microorganisms enables **many copies** of a transferred gene to be made – the gene is **cloned**. It also enables a large amount of the gene product to be obtained.

Obtaining genes

Genes are removed from the DNA of one organism using one or more **restriction endonuclease enzymes**.

- Each enzyme **cuts** the DNA at a **specific base sequence**.
- Restriction enzymes which produce 'sticky ends' are often used.

Transfer of genes

A **vector** is usually required to transfer the 'foreign' gene into bacterial cells.

- **Plasmids** are often used as vectors – small, circular pieces of DNA found in some bacteria.
- The plasmid is cut using the **same restriction endonuclease** enzyme used to remove the foreign gene.
- The plasmid DNA and 'foreign' DNA (gene) are then joined together using a **ligase** enzyme that fits together 'sticky ends'.
- The plasmid and the foreign DNA are referred to as **recombinant plasmid.**
- Plasmid vectors are added to a culture of bacteria and some take up the recombinant plasmid (a process called transformation).

Obtaining gene products

- Bacteria taking up the transferred gene replicate it during cell division, producing a clone of genetically engineered bacteria.
- The bacterial cells are cultured and produce the gene product in large amounts, e.g. insulin, blood clotting factors.

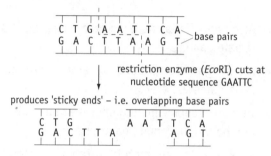

restriction enzyme (*Eco*RI) cuts at nucleotide sequence GAATTC

produces 'sticky ends' – i.e. overlapping base pairs

Action of a restriction endonuclease enzyme

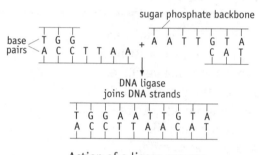

Action of a ligase enzyme

✓ *Quick check 1, 2*

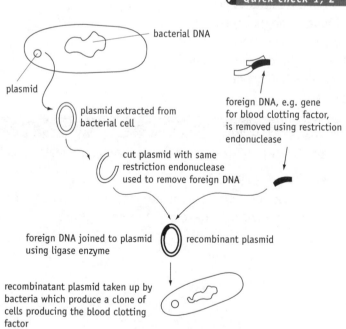

Example of genetic engineering

Genetic markers in plasmids

These are genes already present in a plasmid that allow the detection of bacteria which have taken in the plasmid and the foreign DNA/genes it carries. Markers include genes that give **antibiotic resistance**, e.g. ampicillin resistance.

- Initially the recombinant plasmids are added to bacteria on solid nutrient agar medium (**master plate**).

- After colonies have grown they are transferred to **replica plates** containing the antibiotic ampicillin.

- Only bacteria containing the recombinant plasmid with the ampicillin resistance gene (genetic marker) are not destroyed by the antibiotic.

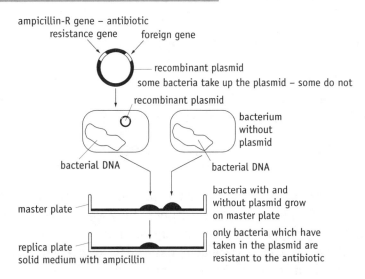

Introduction and detection of recombinant plasmids

✓ **Quick check 3**

Large-scale culturing

- Bacteria containing the recombinant plasmid can be cultured on a large scale in **industrial fermenters**.

- Chemicals produced using large-scale culturing include **enzymes**, **antibiotics** and **hormones**.

- The fermenters are **sterilised** before adding the liquid medium and genetically engineered bacteria.

- Temperature and pH are controlled to provide optimum growth conditions.

- The bacteria divide and produce the gene product, e.g. insulin, which can be extracted.

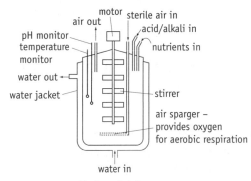

Industrial fermenter

✓ **Quick check 4**

? *Quick check questions*

1 Explain the function of the following enzymes in genetic engineering:
(i) restriction endonuclease; (ii) ligase.

2 What are plasmids?

3 Explain how genetic markers enable detection of genetically engineered bacteria.

4 Name two gene products produced by large-scale culturing.

> Fermenters can be sterilised using steam or disinfectant. This is done to make sure that only the genetically engineered microorganisms grow in the fermenter.

Genetically modified animals

Genetic engineering can be used to produce genetically engineered animals.
Transgenic organisms have a foreign gene added from a different species and can be used to produce substances useful in treating human diseases.

Genetically engineered sheep

The human gene for the production of the protein **alpha-1-antitrypsin** has been inserted into sheep. Alpha-1-antitrypsin is used in the treatment of **cystic fibrosis** and a respiratory disease called **emphysema**.

- The alpha-1-antitrypsin gene was combined to a gene coding for milk production so that it is only active in the mammary glands.
- The combined genes were injected into fertilised eggs which were implanted into recipient females.
- The gene passed on to the offspring and successive generations.
- Ewes (female sheep) produce the protein in their milk, acting as biological factories.
- The protein is expensive to produce in the laboratory and can be extracted at lower cost from the milk.
- Alpha-1-antitrypsin inhibits an enzyme that damages lung tissue.

Evaluation of genetic engineering

Debate about risks of genetic engineering (particularly in medicine and food production) involves:

- possible transfer of foreign genes to non-target organisms (including humans);
- irreversible nature of the process, with no certainty of economic benefits;
- unknown ecological and evolutionary consequences;
- development of resistant species, e.g. antibiotic-resistant bacteria;
- effects of eating genetically engineered food containing foreign proteins;
- accidental transfer of unwanted genes and diseases by the vector, e.g. virus;
- ethical considerations with regard to altering genetic make-up of animals.

> ▶ Newspapers and science periodicals have up-to-date information on the debate concerning genetic engineering.

> ✓ Quick check 1, 2

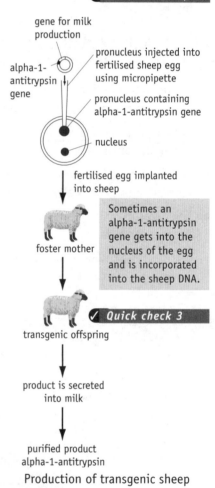

Sometimes an alpha-1-antitrypsin gene gets into the nucleus of the egg and is incorporated into the sheep DNA.

✓ Quick check 3

Production of transgenic sheep

? *Quick check questions*

1 What is a transgenic organism?

2 Name a useful substance produced from a transgenic animal.

3 Describe two possible risks associated with the use of genetic engineering.

50

Cloning

A **clone** is a group of **genetically identical** organisms produced from one parent by asexual reproduction – involving **mitosis**. Plants can be cloned by **vegetative propagation** and animals by the **splitting of embryos**.

Vegetative propagation

- Some plant cells can divide and form new cells of all types (meristems).
- If a meristem is removed with part of a plant, it can grow missing parts.
- Humans use this to grow numbers of genetically identical plants from separated stems (cuttings), buds (grafting), leaves, or roots.

Plant tissue culturing or micropropagation

- Many small pieces (**explants**) of growing plant tissue, e.g. shoot tips, are cut from a valuable plant.
- Their surfaces are sterilised and they are transferred to sterile culture medium containing nutrients and plant growth hormones.
- Shoots and roots develop and plantlets grow which can be planted in soil.

Cloning from embryos

The cells of embryos contain all the genetic information required to form a complete organism and are still undifferentiated, i.e. they are not yet specialised.
- The developing **embryo** consists of a small ball of **genetically identical cells**.
- This can be deliberately split in two – or into individual cells.
- Under appropriate conditions, each cell grows by mitosis into a new embryo.
- This process can be repeated many times to form a large number of genetically identical embryos – clones.
- In mammals, the embryos would have to be transferred into recipient (surrogate) mothers for full development.

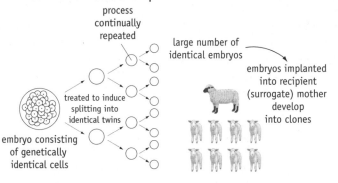

Cloning from embryos

process continually repeated

large number of identical embryos

treated to induce splitting into identical twins

embryos implanted into recipient (surrogate) mother develop into clones

embryo consisting of genetically identical cells

> New, valuable types of roses are produced by selective breeding programmes – using sexual reproduction. Large numbers of new plants are produced by taking buds and grafting them onto the cut-off stems of other rose types.

✓ Quick check 1, 2

> Micropropagation enables virus-free plantlets to be commercially grown

✓ Quick check 3

disease-free stock plant

1 remove small piece (explant) of growing tissue e.g. shoot tip

2 surfaces of explant sterilised with e.g. sodium hypochlorite

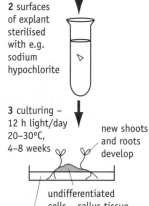

3 culturing – 12 h light/day 20–30°C, 4–8 weeks

new shoots and roots develop

undifferentiated cells – callus tissue

culture medium with nutrients and plant hormones

4 plants grown in greenhouses

5 some cells placed on new culture medium and process repeated to maintain plant cell culture

Stages involved in tissue culturing

? ## Quick check questions

1. What is: (i) a clone; (ii) an explant?
2. Explain why explants must be sterilised before culturing.
3. Draw a flow chart to show the process of cloning using an immature embryo.

Gene therapy and cystic fibrosis

Gene therapy is intended to treat a **genetic disease** by introducing copies of a healthy gene to replace the function of a defective gene.

This process includes:

- identifying the gene causing the disease;
- obtaining and cloning healthy copies of this gene;
- transferring healthy genes into the patient (e.g. by using a **vector**);
- ensuring that genes reach target cells and function.

✓ *Quick check 1*

Cystic fibrosis

Cystic fibrosis is an inherited disorder caused by a defective gene. This gene codes for a chloride channel protein – **cystic fibrosis transmembrane regulator protein** (**CFTR**) – controlling the movement of **chloride ions** in and out of cells.

- In cystic fibrosis a (recessive) **gene mutation** results in a **missing amino acid** (phenylalanine) in the CFTR protein.
- This defective protein does not function normally, and negatively charged chloride ions remain in the cell – high concentrations of positively charged sodium ions balance the negative charge.
- High ion concentrations in the cell prevent water leaving by osmosis.
- This leads to production of abnormally **thick and sticky mucus** by **epithelial cells**.

> ▶ Carriers of cystic fibrosis have one mutant gene and one healthy gene and are not affected by the disease. Two carrier parents can produce a child inheriting two mutant genes and therefore having cystic fibrosis.

✓ *Quick check 2*

Implications of cystic fibrosis

- In the lungs accumulation of thick mucus:
  - narrows air passages, causing **breathing difficulties**;
  - in alveoli **increases diffusion distance** and reduces surface area for gaseous exchange;
  - traps microorganisms – causing repeated lung infections with gradual destruction of alveoli.

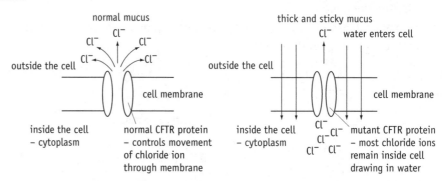

Function of CFTR

- Digestive problems arise as thick mucus blocks the pancreatic duct and intestine, resulting in:
  - small quantities of enzymes – slowing digestion;
  - increased diffusion distance for the absorption of digested food.

- Physiotherapy helps to remove mucus to aid breathing and enzyme supplements improve digestion.

- Males are almost always infertile, females frequently so, because mucus blocks ducts in the reproductive systems.

- Life expectancy is 20–30 years due to lung damage.

✓ Quick check 3

Gene therapy and cystic fibrosis

The healthy gene for the production of CFTR has been identified and cloned. Clinical trials using gene therapy to treat cystic fibrosis have been partially successful in altering the epithelial cells in the lungs.

Vectors

Liposomes and viruses have been used as **vectors** to transfer healthy CFTR genes into epithelial cells.

- Vectors are mixed in a liquid and inhaled into respiratory passages as an **aerosol**.

- Treatment is repeated every few weeks, as epithelial cells die and only some cells take up the vector/gene.

Liposomes:

- are small, spherical vesicles made of a membrane of **lipid molecules**;

- the membrane also contains sugars recognised by receptor proteins on the cell surface membrane of target cells;

- fuse with the cell membrane allowing the CFTR gene to enter the cell, attach to the cell's DNA, and code for normal membrane protein.

> DNA is far too large and has the wrong sort of chemical composition to pass through membranes on its own – that's why it normally stays in the nucleus. So a vector is needed to get it into cells.

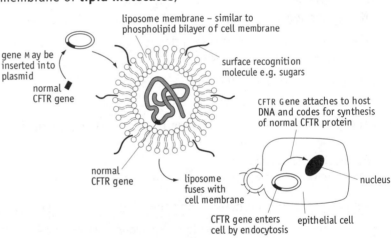

Use of liposomes in gene therapy

Viruses:

- CFTR gene is inserted into harmless viruses;

- the virus enters the epithelial cells of the lungs and the CFTR gene attaches to cell's DNA and codes for the normal membrane protein.

✓ Quick check 4

? Quick check questions

1 What is gene therapy?

2 What does a CFTR gene code for?

3 Explain why an individual affected by cystic fibrosis has breathing difficulties.

4 Name the two vectors used in the treatment of cystic fibrosis by gene therapy.

DNA technology

Scientists can now analyse small samples of DNA.

The polymerase chain reaction (PCR)

This produces large quantities of identical DNA from a small sample – **DNA amplification**. Small samples of DNA are obtained at crime scenes, from an extinct organism or when a gene is isolated for gene therapy.

> ▶ DNA polymerase is obtained from bacteria living in volcanic vents and is not denatured at the high temperatures used in PCR.

Procedure for PCR

- Separate the two polynucleotide strands of DNA by **heating** at 95°C for 5 minutes.

- Add short RNA strands (**primers**), DNA nucleotides and heat-stable **DNA polymerase** enzyme.

- The primers provide a starting sequence for DNA replication.

- **Cool** to 60–70°C for 2 minutes – primers attach to single-stranded DNA.

- **DNA polymerase** adds DNA nucleotides to produce **complementary strands**.

- Repeat cycle of heating and cooling to produce large quantities of DNA.

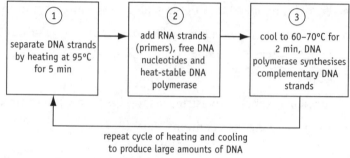

repeat cycle of heating and cooling
to produce large amounts of DNA

Polymerase chain reaction

✓ *Quick check 1, 2*

DNA sequencing

The **Sanger procedure** for determining the sequence of nucleotides in DNA.

✓ *Quick check 3*

- Many copies of the DNA molecule to be sequenced are made by **PCR**.
- These are separated into single polynucleotide strands.
- These are primed for DNA replication and divided into four samples.

Each sample is mixed with a reaction mixture containing:

- **DNA polymerase** enzyme and the **nucleotides** A, T, C and G – to make new **complementary** strands;

- a **percentage** of **one** of the nucleotides with each sample is **chemically altered** to stop formation of a new complementary strand when it is added;

- **radioactively labelled nucleotides** – identify new complementary strands.

Take the example of the reaction with chemically altered thymine.

- As new complementary strands are formed, thymine will base pair with the first adenine on the original strands.

- In some cases the thymine is the altered form and DNA replication stops – giving a short **DNA fragment** of a certain size.

- In most cases replication continues to the next adenine where altered thymine sometimes stops replication – giving larger DNA fragments.

- This process continues to the end of the sample strand.

- The same process occurs with the other altered nucleotides with the other samples – giving a whole range of sizes of DNA fragments, depending on where a particular altered nucleotide stopped replication.

✓ *Quick check 4*

Separation and identification

- The DNA fragments are separated by **gel electrophoresis**.

- Fragments formed using the four altered nucleotides are placed in **separate** wells at the top of the gel.

- **Smaller** DNA fragments move **further** in the gel when an electrical potential is applied.

- After electrophoresis, autoradiography is used to show the position of the radioactively labelled fragments (new complementary strands) which appear as dark bands on X-ray film.

- The **smallest** DNA fragments move the furthest, and were formed when a **specific altered nucleotide** formed a base pair with the **first** nucleotide of the original DNA strand.

- The fragments which moved the next furthest were formed when a specific altered nucleotide formed a base pair with the second nucleotide – and so on down the gel.

(a)

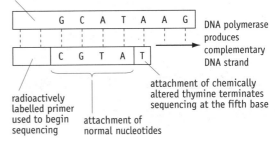

sample DNA with unknown nucleotide sequence acts as a template – millions of DNA molecules are present in the reaction tubes

DNA polymerase produces complementary DNA strand

attachment of chemically altered thymine terminates sequencing at the fifth base

radioactively labelled primer used to begin sequencing

attachment of normal nucleotides

size of DNA fragment will vary depending on where terminating nucleotide attaches

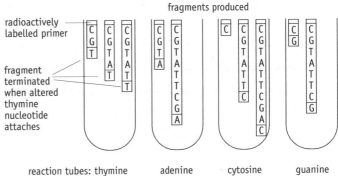

Formation of DNA fragments

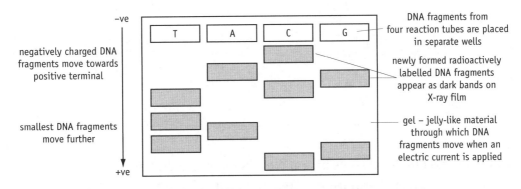

Gel electrophoresis

? Quick check questions

1 What function do primers perform in the polymerase chain reaction?

2 How many DNA molecules could be produced from a single DNA molecule if four complete cycles of the polymerase chain reaction took place?

3 Explain the function of chemically altered nucleotides in the Sanger process.

4 How are DNA fragments separated and identified during DNA sequencing?

Module 2: end-of-module questions

1 The diagram shows part of a DNA molecule.

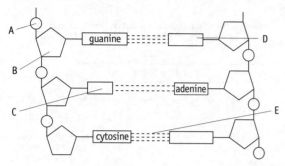

a What do the letters A, B, C, D and E shown on the diagram represent? [5]

b Give two ways in which the structure of DNA differs from RNA. [2]

2 a The diagram shows the sequence of bases in part of an mRNA molecule.

C G U A C C U G U A A U G A U C G U C C U

i Name the process by which mRNA is formed in the nucleus. [1]

ii How many different amino acids could this piece of mRNA code for? [1]

iii Give the DNA sequence which would be complementary to the
first four bases in this piece of mRNA. [1]

b The diagram shows the structure of tRNA.

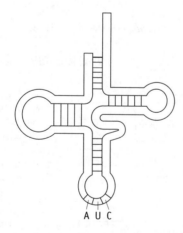

A U C

i What term is used to describe the base triplet shown on the diagram? [1]

ii Give two ways in which the structure of tRNA differs from mRNA. [2]

iii Describe the role of tRNA in protein synthesis. [3]

3 a Describe the semi-conservative mechanism of DNA replication. [6]

b During DNA replication, mutations may arise as a result of mutagenic
agents. Name two types of mutagenic agent. [2]

c Explain why the vast majority of gene mutations are harmful to
organisms. [2]

4 Lysozyme is an enzyme consisting of 129 amino acids.

   a   How many bases would code for this number of amino acids?   [1]

   b   Explain how a mutation in the gene coding for this enzyme may result
          in the production of a non-functional enzyme.   [4]

   c   Suggest why gene mutations involving the deletion of a single base are
          usually more harmful than mutations involving single base substitutions.   [3]

5 The diagrams A, B and C represent three stages of mitosis.

   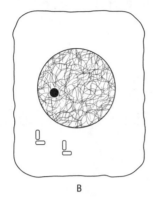   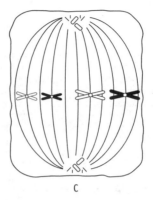

          A   B   C

   a   Name the stages of mitosis represented by A, B and C.   [3]

   b   The table shows the number of cells observed at each stage of mitosis on a
          slide of a plant root.

Stage of mitosis	Number of cells
Interphase	90
Prophase	71
Metaphase	34
Anaphase	15
Telophase	65

   i Use the data to calculate the percentage of time spent in interphase.
     Show your working.

   ii Suggest two factors which may affect the rate of mitosis in plant roots.

   iii Give two processes that occur during interphase that are necessary for
       nuclear division to take place.   [2]

6 The number of chromosomes in a skin cell of a mammal was found to be 44,
   consisting of 22 homologous pairs.

   a   What is the: (i) haploid number; (ii) diploid number in this mammal?   [2]

   b   Explain what is meant by a homologous pair of chromosomes.   [2]

   c   Explain how sexual reproduction in a mammal enables a constant
          chromosome number to be maintained from generation to generation.   [3]

   d   Briefly describe the cloning of animals by splitting apart the cells of
          developing embryos.   [3]

7 Different strains of the bacterium *Pseudomonas* are capable of breaking down
   various chemicals present in oil. Genetic engineering has enabled a '*super
   Pseudomonas*' to be produced by transferring the relevant genes into one
   particular strain of the bacterium.

a Describe how genetic engineering techniques could have been used to (i) remove the relevant genes from the DNA of the *Pseudomonas* strains; (ii) insert these genes into the DNA of the *'super Pseudomonas'*. [4]

b Suggest why the use of *'super Pseudomonas'* to remove oil slicks in the North Sea has been of limited success. [2]

c Describe how genetic markers would enable *'super Pseudomonas'* to be detected amongst the other strains of the bacterium. [4]

8 The diagram shows the stages involved in the polymerase chain reaction.

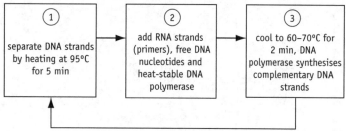

repeat cycle of heating and cooling
to produce large amounts of DNA

a   i Explain why the DNA polymerase enzyme used in this process must be heat-stable. [2]

ii Describe how DNA polymerase synthesises complementary strands. [2]

b After each cycle of the polymerase chain reaction, the number of DNA molecules produced is doubled. How many DNA molecules would be produced from a single DNA molecule after eight complete cycles?

c The polymerase chain reaction enables large amounts of DNA to be produced from a small sample. Suggest two ways in which this may be useful to scientists. [2]

9 Genetically engineered microorganisms can be cultured on a large scale in industrial fermenters to produce useful substances.

a Name two useful products that can be produced by this process. [2]

b Suggest why it is essential to sterilise nutrients before adding them to a fermenter. [1]

c Suggest three advantages of using microorganisms to produce industrial products rather than by chemical processes. [3]

10 In cystic fibrosis a recessive gene mutation leads to the production of a defective protein, CFTR. Affected individuals secrete thick and sticky mucus.

a Explain how the production of the defective protein CFTR leads to the secretion of thick, sticky mucus. [3]

b Explain how digestive problems arise due to the secretion of this thick mucus. [3]

c Describe how gene therapy is being used for the treatment of cystic fibrosis. [4]

Module 3(a): Physiology and Transport

This module is broken down into four topics: Transport systems; The control of breathing and heartbeat; Energy and exercise; and The transport of substances in plants.

Transport systems
- In large organisms diffusion alone is not fast enough to keep cells supplied with the substances they need and to get rid of waste products.
- Transport systems have evolved to carry substances between exchange surfaces of organisms and their tissues, using mass movements of water or gases.
- The final stages in the exchange of substances with the environment and with cells still involve diffusion, osmosis and active transport: all processes involved in crossing cell membranes.
- Mass transport systems maintain the concentration gradients important to diffusion.
- In humans, the mass transport liquid is blood. This flows inside blood vessels, pushed along in one direction by the pumping action of the heart. Oxygen for respiration is carried in the blood, bound to haemoglobin in red blood cells. It is carried from the exchange surfaces of the lungs to the tissues, and waste carbon dioxide is carried back to the lungs to be excreted. Exchange of substances between the blood and the tissues takes place in the capillaries.

The control of breathing and heartbeat
- The rates of heartbeat and breathing have to change, depending on the demands of the body.
- During exercise muscles use more energy, which is supplied by an increase in the rate of respiration. This higher rate of respiration demands a faster supply of oxygen and respiratory substrates and the rapid removal of carbon dioxide. Heartbeat and breathing rates increase in response to this demand. These responses involve receptors in blood vessels and the medulla in the brain.

Energy and exercise
- Muscles use a variety of respiratory substrates, depending upon the size and length of the effort required of them.
- In vigorous exercise, anaerobic respiration takes place which releases much less energy as ATP than aerobic respiration. It also produces lactic acid which is transported to the liver in the blood and converted into glucose.

The transport of substances in plants
Plants have vascular tissues called xylem and phloem, contained in vascular bundles.

The xylem transports water and dissolved mineral ions from the roots to the stem.
- Water enters the xylem in the roots from soil water.
- The uptake of water depends on active transport of mineral ions and osmosis.
- Movement of water up the stem is largely due to the transpiration stream, powered by heat energy supplied by the sun.
- Translocation is the movement of sugars and amino acids in the phloem from sources of these substances to places where they are used. This depends on active transport of sugars, osmosis and mass flows of water.
- The movement of ions and sugars in the vascular tissues can be followed using radioactive isotopes as tracers.

Transport systems – the heart

Larger organisms, like humans and flowering plants, have a small surface area to volume ratio. Specialised exchange systems are needed to obtain oxygen, water and nutrients from the environment and to excrete wastes like carbon dioxide and urea. Exchange surfaces have large areas which are thin and moist for rapid diffusion in solution.

- In larger organisms **exchange systems** work with **transport systems.**
- Transport systems move substances to and from exchange surfaces.
- This prevents build-up of substances at the exchange surface and maintains **concentration gradients.**
- **Mass transport** involves the movement of volumes of water (or gas) carrying substances over **long distances** through a transport system – the **bulk movement** of substances.
- For example, blood plasma is mainly water, carrying glucose, amino acids and carbon dioxide (as carbonate ions) in solution.

Mammalian heart

The blood vascular system circulates blood in blood vessels. Blood is pushed through the vessels by the pumping action of the heart. The heart has right and left sides, each having an atrium and a ventricle.

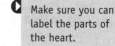
Make sure you can label the parts of the heart.

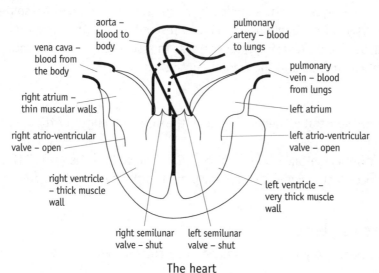

The heart

- **Atria** – receive low-pressure blood returning to the heart in **veins.**
- **Ventricles** – thick muscle to pump blood at high pressure into **arteries.**
- Right atrium – receives **deoxygenated** blood from the body, except the lungs.
- Right ventricle – pumps deoxygenated blood to the lungs (in pulmonary arteries).
- Left atrium – receives **oxygenated** blood from the lungs.
- Left ventricle – pumps oxygenated blood to the rest of the body (in the aorta).

- The left ventricle has a much thicker wall than the right ventricle, because it pumps blood to much more of the body.
- Blood flows in **one direction** through the heart and blood vessels.
- **Atrioventricular valves** open to allow blood into the ventricles as they relax.
- They close as the ventricles contract and blood has to move into the arteries.
- **Semilunar valves** open to allow blood into the pulmonary artery and aorta as the ventricles contract.
- They close as the ventricles relax, preventing backflow into the ventricles.

► Make sure you can explain how the order of beating of the atria and ventricles and the opening and closing of the valves make blood flow in one direction through the heart.

✓ *Quick check 1*

Cardiac cycle

This is the sequence of contraction and relaxation of the heart chambers, and opening and closing of valves, during one heart beat. Volumes and pressures in the heart chambers change during the cycle – and so does the pressure in the aorta.

✓ *Quick check 2*

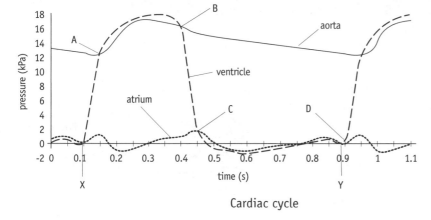

Cardiac cycle

► Make sure you understand how changes in the pressure of the blood open and close valves, and that you can interpret graphs like the one on the left.

- Time A – left ventricle contracting (the pressure inside increasing until it is **greater** than in the aorta).
- The semilunar valve opens and the ventricle empties – blood flows into the aorta.
- Time B – ventricle relaxing and elasticity causing its volume to increase, reducing the pressure inside to **below** that in the aorta.
- The semilunar valve closes, preventing back-flow of blood into the ventricle.
- Time C – pressure inside the ventricle falling below that in the atrium. The atrium contracting, producing a small pressure.
- The atrioventricular valve opens, allowing blood to flow into the ventricle.
- Time D – ventricle contracting again and the pressure inside rising above that in the atrium, making the atrioventricular valve close and preventing back-flow of blood.
- X and Y mark one cardiac cycle; the graph starts to repeat itself after Y.

✓ *Quick check 3*

? Quick check questions.

1 Explain why blood flows in only one direction through the heart.
2 Use the data in the graph to calculate the rate at which the heart is beating.
3 Suggest why the flow of blood around the body is often described as 'double circulation'.

Transport systems – blood vessels

Blood vessels

Blood flows **away** from the heart in **arteries**, which branch into smaller arteries and arterioles. Arterioles branch into **capillary beds**, where **exchange of substances** with body tissues takes place. Capillaries merge into venules, then **veins**, which carry blood **towards** the heart. Arteries carry blood rapidly under high pressure. As the blood flows into arterioles and capillary beds, resistance to the flow of blood increases, causing blood pressure and rate of flow to fall. Blood flows slowly at low pressure in venules and veins back to the heart.

Arteries

Each heart beat sends a surge of blood under **high pressure** along arteries, pushing the blood forwards. The thick artery wall resists pressure and recoils.

- **Elastin** fibres have **elastic** properties.
- Endothelial cells form a layer covering the inside of all blood vessels.
- The lining layer is highly folded to allow for expansion of the artery with each surge of blood.

fibrous layer – to prevent splitting of artery wall due to high blood pressure

smooth muscle layer – thick muscle layer and elastin fibres, to resist expansion of artery wall and make it recoil to squeeze blood and maintain high pressure

lining layer – connective tissue and elastin fibres, with a lining layer of endothelial cells in contact with the blood

small lumen, carrying blood at high speed and pressure

Artery

Arterioles

Arterioles do not have to stand the very high pressure found in main arteries.

- The wall of an arteriole consists of an endothelial layer and **smooth muscle**.
- The smooth muscle can contract or relax, to control the flow of blood into the capillaries.
- The smooth muscle is under **nervous control** (sympathetic nervous system).

Veins

Veins carry blood under **low pressure** towards the heart, so their walls do not have to be very thick.

- The lumen is very large – so even at low speed and pressure, blood flows back to the heart at the same rate that it leaves along the arteries.
- The pressure in the veins is not enough to lift blood from the lower body back to the heart.
- Contracting muscles in the legs and body press on the veins and squeeze the blood along.
- Veins have semilunar valves at intervals to make sure that the blood travels in one direction.

fibrous layer – very thin

smooth muscle layer – thin layer of smooth muscle and elastin fibres

lining layer – a layer of endothelial cells

large lumen, carrying blood at low speed and pressure

Vein

 Make sure you can label the layers of the artery, vein and capillary.

✓ *Quick check 1*

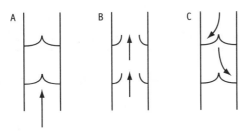

Valves in veins

Make sure you know the function of each type of vessel and whether it carries blood towards or away from the heart.

- In A, blood is pushed up the vein and opens the semilunar valves – B.
- In C, blood tries to flow back down the vein causing the valves to close.

Capillaries

- The walls of the capillaries are one endothelial cell thick, giving a very short pathway for the exchange of substances with the tissues.
- There are very large numbers of capillaries, giving a large surface area for exchange with the tissues.
- No cell in tissues is very far from a capillary, giving short diffusion pathways.

✓ *Quick check 2, 3*

Quick check questions

1 Use the information in the diagrams of an artery and a vein (above) to calculate how many times thicker the wall of the artery is compared with the vein. Assume that the diagrams are drawn to the same scale.

2 Explain how arteries and veins are not adapted for the exchange of substances with the tissues, but capillaries are.

3 Suggest why we can feel a pulse where arteries run near the surface of the body.

Transport systems – exchange

Exchange of materials

Blood is liquid blood plasma, suspended blood cells and substances in solution. Blood plasma is water with: glucose (blood sugar); amino acids and fatty acids; hormones (e.g. insulin, glucagon); urea; mineral ions (sodium, potassium, chloride, bicarbonate [carbon dioxide]); and proteins (e.g. enzymes and antibodies). Substances enter and leave the blood capillaries at exchange surfaces.

Exchange of materials		
Substance(s)	Site of capillaries where it enters blood	Site of capillaries where it leaves blood
Oxygen	Alveoli of lungs	Tissues of body
Carbon dioxide	Body tissues, except lungs	Alveoli of lungs
Glucose, amino acids, fatty acids, mineral ions	Epithelium of villi of small intestine	Tissues of rest of the body
Hormones	(Endocrine) glands	Target organs/tissues
Urea	Liver	Kidney

> ▶ Make sure you know which substances are exchanged at which exchange surface.

- Proteins are secreted into the blood by e.g. liver cells.
- Antibodies are secreted by white blood cells (lymphocytes) directly into the blood plasma.

Loading, transport and unloading of oxygen

Oxygen is carried in **red blood cells**, reversibly bound to **haemoglobin** – a **protein** with **specific binding sites** for four oxygen molecules.

> **Haemoglobin + Oxygen ⇔ Oxyhaemoglobin**

- Blood entering lung capillaries is deoxygenated.
- **Alveoli** contain a high concentration of oxygen, giving a **concentration gradient** for the **diffusion** of oxygen through the wall of the alveolus and capillary into the blood plasma, red cells and haemoglobin.
- Oxygen is carried away by the blood, maintaining a concentration gradient.
- At rest, muscles take little oxygen for respiration.
- During exercise, muscle respiration increases, more oxygen is used and a higher concentration of carbon dioxide is produced.
- This is converted into carbonic acid in the red cells and **lowers the pH** of the blood plasma.
- This causes haemoglobin to release more oxygen to maintain the high rate of respiration.
- The **oxygen–haemoglobin dissociation curve** shows the relationship between the amount of oxygen carried by the blood and the amount of oxygen in the tissues.

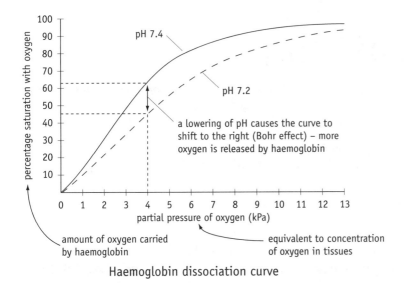

Haemoglobin dissociation curve

a lowering of pH causes the curve to shift to the right (Bohr effect) – more oxygen is released by haemoglobin

amount of oxygen carried by haemoglobin

equivalent to concentration of oxygen in tissues

> Make sure you can explain the links between respiration in muscles during exercise, rises in blood carbon dioxide, and the release of more oxygen to the muscles by haemoglobin.

- In the steep part of the dissociation curve, a small fall in oxygen in the tissues causes a lot of oxygen to dissociate from haemoglobin.

✓ *Quick check 1, 2*

Tissue fluid

Tissue fluid surrounds cells in tissues and is formed from substances which leave the plasma in blood capillaries. Compared to plasma it has: a lower concentration of oxygen and higher concentration of carbon dioxide; and no blood cells, platelets or plasma proteins (they are too large to cross the endothelial cells of the capillary).

- Oxygen, glucose and mineral ions diffuse into the tissue along concentration gradients – maintained by cells constantly using them.
- Carbon dioxide and urea diffuse into the plasma along concentration gradients.
- Not all the water leaving the capillary is reabsorbed.
- **Lymph** forms from surplus tissue fluid draining into lymph capillaries of the **lymphatic system**.
- These capillaries are dead-end vessels which merge to form larger lymph vessels, which return lymph to the blood plasma.
- Lymph is similar to tissue fluid but can contain fats, more protein and white blood cells.

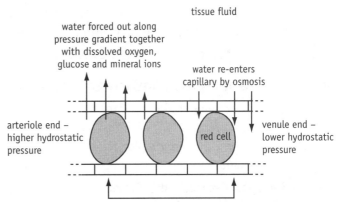

Exchange of water in a capillary

✓ *Quick check 3*

? Quick check questions

1 Find the amount of oxygen dissociation caused by a fall in pH of 0.2 at a partial pressure of oxygen of 4 kPa.

2 Carbon monoxide in cigarette smoke has a greater affinity than oxygen for haemoglobin. Suggest why smoking makes exercise harder to maintain.

3 Explain how an oxygen molecule in an alveolus reaches a muscle cell in the leg.

Control of breathing

During exercise, **muscular activity** and rate of **respiration** increase. The rate of heartbeat and the rate and depth of breathing/ventilation increase, increasing uptake of oxygen and excretion of carbon dioxide. After exercise, heartbeat and breathing return to normal.

Control of ventilation

Breathing is controlled by the **medulla** in the brain, containing a breathing centre, divided into **inspiratory/inhalation** and **expiratory/exhalation** centres.

✓ Quick check 1

- The inspiratory centre causes intercostal and diaphragm muscles (effectors) to contract, inhibits the expiratory centre, and we breathe in.
- As the lungs inflate, **stretch receptors** send **nerve impulses** to the medulla, **inhibiting** the inspiratory centre (Hering–Breuer reflex).
- The expiratory centre is no longer inhibited, and we exhale.
- As the lungs deflate, stretch receptors become inactive, the inspiratory centre becomes active and the expiratory centre is inhibited.
- These events produce a rhythmical breathing action.

> You can use the flow-chart to help you remember facts but it isn't an 'explanation' on its own. Try to explain in writing what the chart shows.

Responses to increased muscular activity

Increased muscle activity produces more carbon dioxide from respiration, which is converted to carbonic acid in red blood cells and lowers blood pH.

- **Chemoreceptors** stimulated by lower pH are found in the **aortic body**, **carotid body** (in the aorta and carotid arteries), and **medulla.**
- Nerve impulses travel to the medulla, leading to increases in the rate and depth of breathing and rate of excretion of carbon dioxide.
- After exercise, the rate of breathing remains high until the concentration of carbon dioxide in the blood falls to normal.

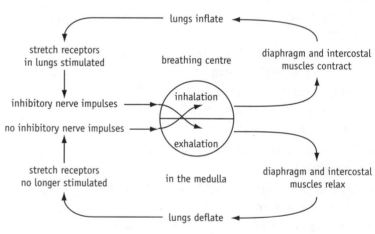

Control of ventilation

✓ Quick check 2

> Make sure you can use the terms used here to explain why breathing rate increases during exercise – simpler terms won't do.

? Quick check questions

1 Explain how the normal rate of breathing is controlled.
2 During vigorous exercise, lactic acid is produced by muscles and released into the blood. Suggest how this could affect the rate of breathing.

Control of heartbeat

The heart beats with its own rhythm. Heart rate increases during muscular activity, increasing the flow of blood to muscles, bringing more oxygen and glucose, and getting rid of more carbon dioxide.

> There are **two** nodes involved in the heart beat.

Sinoatrial node, atrioventricular node and heartbeat

- The **sinoatrial node** (SA node) is a patch of modified muscle cells in the wall of the **right atrium** that produces regular bursts of electrical impulses.
- These impulses spread rapidly through the walls of the right and left atrium, causing them to contract together.
- The impulses reach the **atrioventricular node** (AV node).
- There is a **delay** of 0.15 seconds before the AV node reacts, so that the ventricles contract after the atria.
- Impulses from the AV node travel rapidly through the **bundle of His** and into branches to all parts of the ventricles.
- The ventricles contract, starting at the bottom, to push the blood up and out into the arteries.

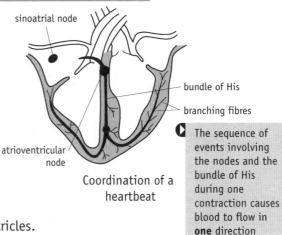

Coordination of a heartbeat

> The sequence of events involving the nodes and the bundle of His during one contraction causes blood to flow in **one** direction through the heart.

✓ **Quick check 1**

Changes in rate of heartbeat with exercise

- Heart rate is controlled by a **cardiac centre** in the **medulla** of the brain, divided into a cardioaccelerator centre and a cardioinhibitor centre.
- Contracting muscles press on veins, force blood towards the heart, and cause greater filling of the ventricles, making the heart beat faster and stronger.
- If blood pressure rises too far above normal, **pressure receptors** in the aorta and carotid artery send nerve impulses to the cardioinhibitor centre.
- This centre sends inhibitory nerve impulses to the cardioaccelerator centre and the SA node – preventing the heart beating too fast.

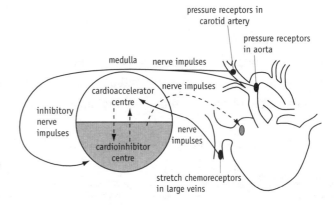

Changes in heart rate with increase in muscular activity

✓ **Quick check 2**

- Stimulation of chemoreceptors sensitive to pH that leads to an increase in rate of breathing also leads to an increase in heart rate.

> Don't confuse how heart beat is regulated with the control of breathing!

? Quick check questions

1 Explain the importance of the time delay before electrical impulses from the sinoatrial node reach the atrioventricular node, to the flow of blood from the atrium to the ventricle.

2 Suggest why it is important to prevent the heart beating too fast.

Energy sources

During exercise, muscular cells/fibres contract more and more strongly. This requires a greater supply of energy. The immediate source of energy for **muscle contraction** is **ATP** from **respiration**. The process of respiration needs **respiratory substrates**. There is a tendency to think only about glucose as the respiratory substrate, but there are a number of possible inputs to respiration.

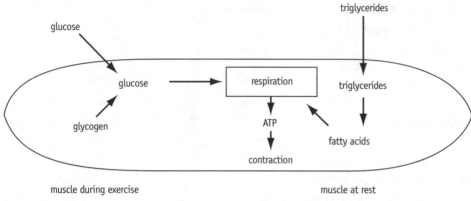

Energy sources for muscles

> Don't fall into the trap of thinking that muscles only use glucose in respiration.

- Resting muscles use **fatty acids** obtained from **triglycerides** as their main respiratory substrate.
- During exercise the muscles switch mainly to **glucose** as the respiratory substrate.
- Muscles contain a lot of **glycogen**, a polymer of glucose, which can be converted to glucose for respiration.
- **Aerobic** respiration produces **38** molecules of ATP per molecule of glucose.
- Aerobic respiration produces carbon dioxide and water as waste products.
- **Anaerobic** respiration produces **two** molecules of ATP per molecule of glucose used.
- Anaerobic respiration produces pyruvate which is then converted to **lactate**.

✓ *Quick check 1*

Muscle fatigue

During prolonged, heavy exercise, aerobic respiration cannot supply all of the demand for ATP by the muscles. This is partly because the blood cannot supply oxygen quickly enough to the muscles.

- The muscles produce a lot of **lactate** by anaerobic respiration.
- The lactate diffuses out into the blood and decreases the pH of the blood.
- The build-up of toxic lactate and other waste products in the muscles causes **muscle fatigue**, pain and tiredness.
- The lactate travels to the **liver**, where it is converted back to pyruvate and then glucose.

> Try not to get mixed up between aerobic and anaerobic respiration.

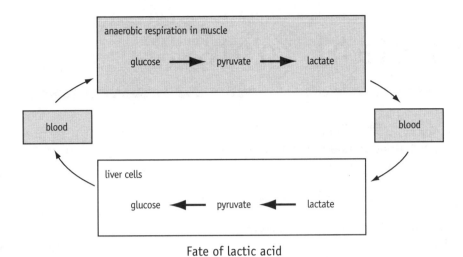

Fate of lactic acid

- This glucose can then return to the muscles in the blood and be used, or converted to glycogen.

- The conversion of lactate back to glucose requires **oxygen**.

- After strenuous exercise, the body/muscles has an oxygen debt until all of the lactate has been converted back.

- This explains why you keep panting/breathing hard for some time after you finish a run – to take in the oxygen needed to convert lactate to glucose.

> ▶ Don't forget the job of the liver in breaking down lactate.

✓ Quick check 2

? *Quick check questions*

1 Explain what happens to respiration in muscle cells during prolonged, heavy exercise.

2 Suggest why sprinters run a 100 m race without any need to breathe.

Transport in plant roots

Uptake of water and mineral ions from the soil occurs in plant roots. These substances are transported across the root tissues and up the plant to the leaves.

Root structure

- The piliferous layer has **root hairs** which are thin, permeable, tubular extensions of single **epidermal cells**.

- The cortex consists of parenchyma cells in which starch may be stored.

- The **endodermis** is one cell thick and has impermeable **Casparian strips** made of suberin in its cell walls.

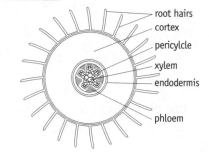

Structure of a primary root

Phloem

The phloem transports **organic molecules** to or from the root, and consists of living cells including sieve elements and companion cells.

- **Sieve elements** possess perforated end walls or **sieve plates**.

- Sieve elements are joined end-to-end to form **sieve tubes**.

- Mature sieve elements have no nucleus and few organelles.

- Each sieve element has a **companion cell** with a nucleus, dense cytoplasm and many mitochondria.

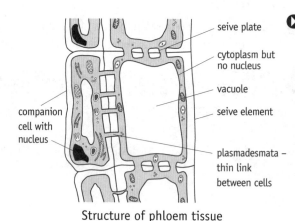

Structure of phloem tissue

> In phloem and xylem there are cells joined end to end to make long 'tubes' through which mass flows of liquid and dissolved substances can take place.

Xylem

The xylem transports **water and mineral ions** from the roots to the leaves. It is a non-living tissue consisting of vessels and tracheids.

- **Vessels** are formed by many xylem cells joining together – their end walls break down to form long, hollow 'tubes'.

- The vessel walls are thickened and strengthened by impermeable lignin.

- Pits in the lignin allow water to pass sideways between vessels.

- **Tracheids** are elongated, lignified single cells with tapering end walls that overlap with adjacent tracheids.

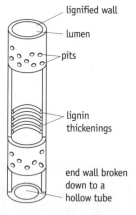

A xylem vessel

> Make sure you can label the parts of the primary root.

✓ *Quick check 1*

Uptake and transport in roots

- Uptake of water and ions is mainly by the root hairs which provide a large surface area for the absorption of water and ions.
- Ions are absorbed into the root by **diffusion** and **active transport**.
- Water uptake is by **osmosis** along a **water potential gradient** – soil water has a higher water potential (less negative) than the root hairs (more negative).
- Water moves along a water potential gradient across the root.
- The water potential gradient is maintained by water continually moving up the xylem and by dissolved ions in the xylem sap.

The movement of water across the root cells can occur via three pathways, the apoplast, symplast and vacuolar.

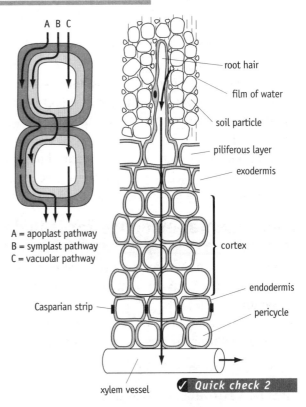

A = apoplast pathway
B = symplast pathway
C = vacuolar pathway

Pathways of transport of water and ions across a root

✓ *Quick check 2*

Apoplast pathway

- This is movement of water through the cellulose **cell walls** of adjacent cells and the small intercellular spaces between them.
- Cell walls are fully permeable across the root, except for the **endodermis**.
- The impermeable **Casparian strip** prevents passage of water and dissolved mineral ions through the apoplast.
- They have to cross cell membranes of endodermal cells into the symplast, allowing control of the movement of water and ions into the xylem.
- Endodermal cells actively pump ions into the xylem.

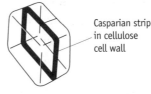

Casparian strip in an endodermal cell

❚ The same three pathways account for the movement of water in the leaf from the xylem tissue, across the leaf mesophyll cells and out through the stomata

Symplast pathway

- This is movement of water, by osmosis, through the interconnecting cytoplasm of adjacent cells.
- The water travels through plasmodesmata – thin strands of protoplasm linking the cytoplasm of adjacent cells.

✓ *Quick check 3, 4*

Vacuolar pathway

- This is the movement of the water through cell vacuoles of adjacent cells.
- The water must cross the cell surface and vacuole membrane.
- The water moves by osmosis along a water potential gradient.

? Quick check questions

1 Name the parts of the root labelled A, B and C on the diagram (right).
2 Explain how a low water potential is maintained in the xylem tissue of a root.
3 Describe how water movement occurs by the symplast pathway in a root.
4 How do endodermal cells control water movement into the xylem?

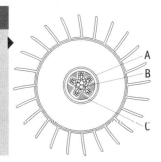

Transport in the phloem

The phloem transports organic substances – a process called **translocation**. Carbohydrate is transported as sucrose; protein as amino acids or amides; and lipids as fatty acid and glycerol. Organic molecules are transported up and down the stem of a plant to:

- respiring cells which do not carry out photosynthesis;
- growing areas such as young leaves, shoot tips and root tips;
- storage areas, e.g. roots and developing fruits.

Experimental evidence for translocation in the phloem

- **Ringing** involves removing a complete ring of phloem (and other tissues) from part of a tree trunk, preventing transport through the phloem but **leaving the xylem**.
- Over time, a swelling develops above the ring due to the build-up of photosynthetic products from the leaves.
- The tree dies because the roots receive no respiratory substrate.

> In the ringing experiments, the xylem was untouched but was unable to transport organic substances to the roots.

> ✓ *Quick check 1, 2*

Use of radioactive tracers

- The radioactive isotope of carbon (^{14}C) can be used to trace the movement of organic substances in plants (and other organisms).
- For example, two plants of the same species and similar stages of growth are taken and the stem of one plant, A, is ringed, while the other plant, B (the control), is left intact.
- A leaf below the ring in plant A, and one at a similar position in plant B, are exposed to radioactively labelled $^{14}CO_2$ (or injected with ^{14}C-labelled glucose) as shown in the diagram.
- The leaf cells use $^{14}CO_2$ to form ^{14}C-labelled glucose during photosynthesis.
- The plants are left for a few hours in sunlight.
- Transport of radioactively labelled photosynthetic products can be traced using autoradiography.
- Each plant is placed under X-ray or photographic film, which is later developed.
- Where the film is over part of the plant containing ^{14}C, ionizing radiation causes the film to react as if exposed to light and a black dot is found.
- Autoradiograms for plants A and B are shown on page 73.
- Plant B shows transport of photosynthetic products to the growing shoot tip, root tip, young leaves and storage areas.
- In A, photosynthetic products remain below the ring where phloem has been removed.
- (A very small amount of radioactivity above the ring is due to a small movement of organic substances between the phloem and xylem tissue.)

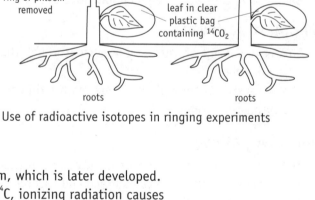

Use of radioactive isotopes in ringing experiments

> Plants and other organisms use radioactive carbon like non-radioactive carbon.

The mass flow hypothesis

In the mass flow hypothesis, translocation in the phloem is due to mass flow of solutions of organic substances along a hydrostatic or turgor pressure gradient.

- Photosynthetic products, e.g. sugars, are produced in the mesophyll cells in the leaves, known as the **source**.
- Sugars are **actively** transported into the sieve tubes – lowering their water potential.
- Water enters by **osmosis** and creates a **high hydrostatic** or **turgor pressure** in the sieve tubes in the leaf.

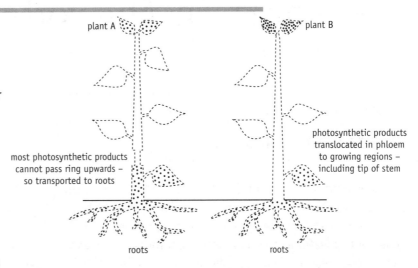

most photosynthetic products cannot pass ring upwards – so transported to roots

photosynthetic products translocated in phloem to growing regions – including tip of stem

roots

roots

Autoradiograms

The water entering the sieve tubes in the leaves is supplied by the xylem.

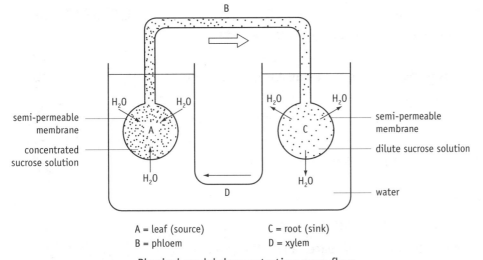

A = leaf (source)
B = phloem
C = root (sink)
D = xylem

Physical model demonstrating mass flow

In the physical model, mass flow eventually stops as sugars are not being continually produced at the 'source' or used up at the 'sink'.

- In the roots, growing areas, etc., **sugars leave the phloem** to be used in respiration, for growth or are stored as insoluble starch – these are **sinks**.
- The water potential of the sieve tubes in the sinks becomes higher, water leaves by osmosis and hydrostatic pressure falls.
- The photosynthetic products are transported along the pressure gradient in the sieve tubes by mass flow.

✓ *Quick check 3*

❓ Quick check questions

1 In what form are carbohydrates transported in the phloem?
2 Explain why fruit on ringed branches grow much larger than normal.
3 Explain what is meant by: (i) a source; (ii) a sink.

Transport in the xylem

In plants, water and mineral ions are transported in the xylem and organic substances in the phloem.

The transpiration stream

The movement of water and dissolved mineral ions from the root hairs to the stomata through the xylem is part of the transpiration stream. There are two main hypotheses for the movement of water and mineral ions from roots to leaves in the xylem: root pressure, and the cohesion–tension hypothesis.

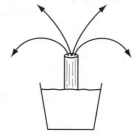

sap exudes from cut end of stem

Root pressure is not found in all plants and is considered to be of secondary importance compared to the cohesion–tension. hypothesis

Root pressure

- Root pressure is a **positive hydrostatic pressure**, seen if a freshly cut root stump continues to exude sap.

- **Active transport of mineral ions** by endodermal cells **lowers the water potential** of the xylem.

- Water moves into the xylem by **osmosis** – raising the hydrostatic pressure in the xylem.

- Respiratory inhibitors, low temperatures and lack of oxygen inhibit this process by stopping active transport.

- Root pressure is not great enough to transport water to the top of trees.

Cohesion–tension hypothesis

- Solar heat energy causes evaporation or **transpiration** of water from leaves.

- Water **evaporates** from mesophyll cells next to air spaces and diffuses out through the stomata into the air.

- The water potential of these mesophyll cells is lower compared to inner mesophyll cells.

- Water moves from adjacent cells by osmosis along this **water potential gradient**.

- Movement of water is by the apoplast, symplast and vacuolar pathways.

- This water potential gradient extends to the xylem vessels, drawing water from the xylem and creating a **tension** in the xylem vessels and 'pulling up' water and dissolved ions.

- Water in the xylem forms a **continuous column** from the leaves to the roots. If the top of the column is 'pulled up', the whole column moves up.

✓ **Quick check 1**

mercury manometer – measures root pressure

Root pressure in a cut stump

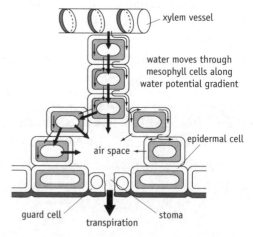

xylem vessel

water moves through mesophyll cells along water potential gradient

epidermal cell

air space

guard cell

transpiration

stoma

Water movement across the leaf

- A continuous water column is maintained by **cohesive** forces and **adhesive** forces – cohesive due to hydrogen bonding between water molecules; adhesive due to attraction of the water molecules to the xylem walls.

- Upward movement of water from the xylem in the roots maintains a water potential gradient across the root cortex cells, for water uptake from the soil via osmosis.

✓ *Quick check 2, 3*

> ◑ Radioactive phosphate ions are often used to trace uptake from the roots to the stem – plants need phosphate for making ATP, DNA, RNA and phospholipids.

Evidence for movement of ions in the xylem

- **Radioactive tracers** can be used to provide evidence that the transport of mineral ions through a plant occurs mainly in the xylem.

- The xylem and phloem in a section of the stem of a plant is separated using a wax cylinder to prevent lateral transport.

- The roots are supplied with radioactive potassium ions, ^{42}K.

- The plant is left for a few hours and the amount of radioactivity in the xylem and phloem tissues in the region of the wax cylinder is then measured and compared.

- The amount of radioactivity in the xylem is considerably greater, indicating that transport of potassium ions occurs in this tissue.

- The small amount of radioactivity in the phloem tissue is due to lateral transport from the xylem in the region where the wax cylinder is not present.

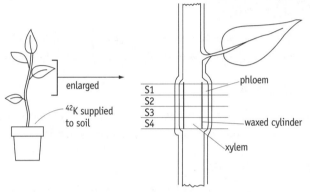

Transport of ^{42}K in plants

| Section number | ppm ^{42}K in tissue | | | |
| | Stripped | | Unstripped (control) | |
	Phloem	Xylem	Phloem	Xylem
S1	0.9	119	49	54
S2	0.3	108	47	70
S3	0.5	112	50	67
S4	0.3	109	48	57

ppm is just a measure of concentration

❓ Quick check questions

1 Briefly describe how you could show that root pressure involves active transport.

2 What is the energy source in the cohesion–tension hypothesis?

3 Explain what is meant by cohesive and adhesive forces in the transport of water in the xylem.

Transpiration

Transpiration is the evaporation of water from a plant's surface, particularly through the stomata. The rate of transpiration is affected by external environmental factors and internal factors related to the structure of the plant.

Environmental factors

Light

- In daylight stomata open, allowing carbon dioxide to enter for photosynthesis.
- More water diffuses out of the leaf – increasing the rate of transpiration.

Temperature

- An increase in temperature gives water molecules more kinetic energy, allowing them to evaporate more easily from mesophyll cells.

Humidity

- An increase in humidity decreases the water potential gradient for the diffusion of water, decreasing the rate of transpiration.

Air movement

- Air movement removes water vapour from the leaf surface, increasing the water potential gradient and transpiration.
- In still air, water vapour builds up around the leaf reducing the water potential gradient and rate of transpiration.

Xerophytes

These are plants that live in habitats where water is in short supply. They have structural adaptations that reduce the rate of transpiration. Many involve structural adaptations of leaf structure such as:

- a **thickened waxy cuticle** reducing evaporation;
- **hairs** on the leaf surface to trap a layer of air – which becomes saturated with water vapour reducing the water potential gradient for water loss;
- **curled leaves** (e.g. marram grass) reduce the surface area for evaporation and increase the humidity in the air around the stomata, reducing transpiration;
- **reduced leaf surface area** (e.g. pine needle) over which transpiration can occur;
- **sunken stomata** (often in epidermal pits) which become saturated with water vapour, reducing the water potential gradient for water loss.

> The surface of a leaf is covered by a waxy cuticle which is there to prevent water loss. It also blocks gaseous exchange – which is why stomata are needed.

> ✓ *Quick check 1*

> Make very sure that you don't confuse translocation (of organic substances in the phloem) and transpiration (of water).

> Plants that live in habitats where the water supply is adequate are known as mesophytes.

> ✓ *Quick check 2, 3*

? Quick check questions

1 Explain why the rate of transpiration is high on a warm, breezy day.

2 What are xerophytes?

3 Describe and explain how two adaptations of leaf structure can reduce the rate of transpiration.

Module 3(a):
end-of-module questions

1 a Explain how the structures of the heart make blood flow in
one direction. [4]

b The graph shows changes in pressure in the left ventricle and
aorta during one cardiac cycle.

i Explain what happens at points X and Y. [4]

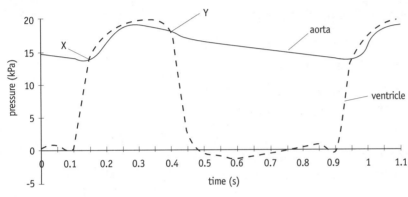

ii Calculate the heart rate in beats per minute. [2]

2 a Explain how the structure of each of the following is adapted to
its function: (i) artery; (ii) capillary. [6]

b The blood in arteries flows very rapidly and under high pressure, but
blood in veins flows slowly and at low pressure. The volume of blood
passing through arteries and veins has to be the same over time,
otherwise blood would accumulate in the tissues. Explain how the
flow rate in the arteries and veins can be the same. [4]

3 a Complete the table. [5]

Substance(s)	Site of capillaries where it enters the blood	Site of capillaries where it leaves the blood
Oxygen		Tissues of rest of body
Carbon dioxide		
Glucose, amino acids, fatty acids and mineral ions	Epithelium of villi of small intestine	
Hormones		Target organs/tissues
Urea		Kidney

b i Explain why most of the water that leaves the blood plasma into the tissue fluid at the arteriole end of a capillary re-enters the plasma towards the venule end of the capillary. [3]

ii Elephantiasis is a disease caused by parasitic worms that invade and block the lymphatic system. Suggest what the effects of this blockage might be. [3]

4 a Explain how the contraction of the atria and ventricles is controlled to make sure that both sides of the heart beat together and that blood flows in the correct direction through the heart. [6]

b The heart beats with its own rhythm but can speed up during muscular activity.

i Explain why the heart needs to speed up. [3]

ii Explain how the speeding up of the heart happens. [6]

5 a Describe the role of the nervous system in the response of the breathing system to increased muscular activity. [6]

b During a race, a sprinter's muscles may release a lot of lactic acid into the blood. Suggest how this may help their muscles to get more oxygen from the blood. [3]

6 a In an answer to an exam question a student wrote that "Muscles use glucose as their source of energy". Is this an accurate statement? [4]

b Explain what happens to the lactate produced in muscles during prolonged heavy exercise. [4]

7 a The diagram shows a cross-section of a root. Identify W, X, Y and Z. [4]

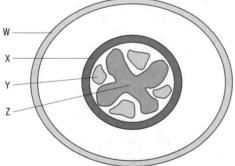

b Explain how water moves into the xylem in the root from the soil. [6]

8 a Explain how a water molecule gets from the xylem in a root to the air outside stomata. [8]

b Describe and explain how three environmental factors affect transpiration. [6]

9 a Explain how a sugar molecule produced by photosynthesis in a leaf cell reaches the cells in a flower growing at the top of a stem. [6]

b In the days of the American 'Wild West', farmers and their families often settled in heavily wooded areas. When they first arrived, they needed to plant and grow enough crops to keep themselves alive until they had time to clear the trees for fields. Unfortunately, the trees did not let enough light reach the ground for crops to grow. The farmers used axes to remove rings of material all round the trunks of trees in an area and then planted crop seeds between the trees. Suggest how ringing the trees allowed the farmers to grow their crops. [5]

10 The experiments shown in the diagram were used to investigate the movements of ions and organic substances in a plant.

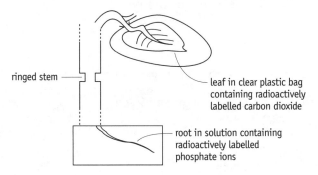

ringed stem

leaf in clear plastic bag containing radioactively labelled carbon dioxide

root in solution containing radioactively labelled phosphate ions

a   Explain the following results:

  i Radioactively labelled phosphate ions reached the leaf. [3]

  ii Radioactively labelled sugars were not found in the root. [3]

b   Describe a method for finding out where the radioactive tracers had moved to in the plant. [3]

c   The plant was exposed to cyanide, a poison that stops respiration. Suggest which tracer's movement this would have the greatest effect on. [6]

Appendix: Exam Tips

Lots of marks are lost by not answering questions as they are set, or not knowing material in the syllabus. You must know the information, terms and examples that are included in the syllabus – the exam board only allows questions that can be answered using syllabus material. Other information will not harm you, but will not be necessary to get a good mark!

Describe

'Describe' means put information into words. The information is usually **given** to you in a table, graph or diagram.

Example: the graph shows the rate of reaction of an enzyme with different concentrations of substrate and how the rate is affected by the addition of a particular concentration of an inhibitor.

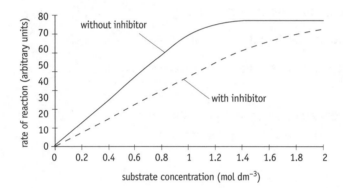

Question

Describe the effect of each of the following on the rate of reaction of the enzyme:

a the concentration of substrate;

b the inhibitor.

Answers

a Between substrate concentrations of 0 and 1 mol dm^{-3}, the rate is directly proportional to the concentration of substrate. The rate reaches a maximum at a substrate concentration of about 1.4 mol dm^{-3}.

b The inhibitor reduces the rate of action of the enzyme at lower concentrations of substrate. Above substrate concentrations of 1.2 mol dm^{-3}, the rate with the inhibitor gets closer and closer to the rate without the inhibitor.

These answers **describe** what you can see on the graph – in reasonable detail.

Explain

'Explain' means that you should 'know' the answer from syllabus material that you have been taught. The following question uses the same graph and information as for 'Describe'.

Question

Explain if the inhibitor is competitive or non-competitive.

Answer

The inhibitor is competitive, because its effect is overcome by increasing the concentration of substrate.

A non-competitive inhibitor would inhibit the enzyme at any concentration of substrate.

Suggest

'Suggest' means that you are unlikely to have seen the material in the question but you should have been taught things from the syllabus that will allow you to answer. The following question applies to an enzyme and its inhibitor.

Question

Two substances, **X** and **Y**, were investigated as possible rat poisons. Both inhibit the same enzyme in an important biological process. The diagrams show the structure of the enzyme, its substrate and the inhibitors. Use the information in the diagrams to suggest which inhibitor would be the best poison.

Answer

Inhibitor Y would be best, because it is non-competitive. It binds to a site other than the active site.

Its effect cannot be overcome by more substrate (unlike inhibitor Y) and so the metabolic pathway is blocked.

OR

Inhibitor X would be best, because it is competitive. It binds to the active site. This reduces/stops the substrate being turned into product (and stops the metabolic pathway).

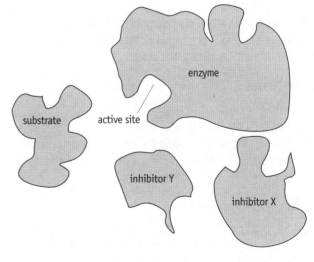

Rat poison

It is not unusual to have alternative answers to this sort of question. You are not expected to have learnt this material – it isn't specifically given in the syllabus. Either answer is a reasonable interpretation of the information.

Answers to quick check questions

Module 1: Core Principles

Carbohydrates

1 Glucose, monosaccharide, reducing; sucrose, disaccharide, non-reducing.
2 Condensation reaction, loss of water.
3 Heat with Benedict's solution; brick-red colour is positive.

Carbohydrates – polysaccharides

1 Glucose in long, straight chains; many hydrogen bonds link chains together; microfibrils provide rigidity.
2 Starch is insoluble; water potential unaffected; helical, compact store; contains many glucose molecules for respiration.
3 Liver or muscle.

Lipids

1 A saturated fatty acid does not possess any double bonds.
2 Heating with acid or alkali; lipase at its optimum temperature.
3 Triglyceride molecule – a glycerol molecule with three fatty acids; phospholipid – one fatty acid is replaced by a phosphate group.

Proteins

1 Nitrogen.
2 See diagram on page 8.
3 The sequence of amino acids in a polypeptide.
4 Denaturation; hydrogen, ionic bonds break; irreversible change of tertiary structure.

Chromatography

1 R_f value = $\dfrac{\text{distance moved by compound from origin}}{\text{distance moved by solvent from origin}}$
2 Locating agents stain colourless spots.
3 2-dimensional chromatography – spot mixture at origin, run chromatogram with a solvent, record R_f values, turn the chromatogram 90°, run second chromatogram using a different solvent, record R_f values.

Water

1 Transport of nutrients / excretory products / secretion of hormones / enzymes.

2 Hydrogen bonds.
3 Water absorbs large amounts of heat energy to heat/cool; minimises changes in temperature in organisms; and aquatic environments.

Cells

1 Three of:

Prokaryotic	Eukaryotic
DNA is circular, no nucleus	DNA is linear, in a nucleus
Diameter of cell 0.5–10 μm	Diameter of cell 10–100 μm
Smaller, 70S ribosomes	Larger, 80S ribosomes
No mitochondria present	Mitochondria present
No Golgi body	Golgi body present
Flagella, lacking microtubules	Flagella have microtubules

2 Advantages – high resolution, high useful magnification; disadvantages – specimens are dead, techniques alter/damage cells, expensive.
3 (i) Prevent action of enzymes which hydrolyse cell components; (ii) prevent osmotic movement of water which cause organelles to burst or shrivel.
4 Nuclei.

Cell structure – organelles

1 Both contain DNA / ribosomes / possess an envelope of two membranes / upper membrane increases surface area for processes.
2 (i) Protein synthesis; (ii) production and transport of lipids; (iii) produce glycoproteins / secrete enzymes / produce lysosomes / form cell walls.

Cell membranes and differentiation

1 Lipid-soluble molecules pass directly through phospholipid bilayer.
2 (i) Transport ions and polar molecules; (ii) allow specific hormones to attach and stimulate cells.
3 (i) Tissues – aggregations of similar cells with a specific function; (ii) organ – consisting of different tissues, with a specific physiological function.

4 Large surface area – provided by microvilli; many mitochondria – providing ATP for active uptake.

Transport across membranes

1 Zero.
2 Water moves into the root hair by osmosis from the higher water potential in the soil.
3 Diffusion – passive process along a concentration gradient; active transport – against a concentration gradient, requires energy.
4 Lower the temperature / reduce the amount of oxygen available.

Gaseous exchange in mammals

1 As size of organism increases, surface area: volume ratio decreases.
2 Large surface area / moist surface / short diffusion pathway.
3 Intercostal muscles contract – move ribcage up and outwards; diaphragm muscles contract and diaphragm flattens; volume of thorax increases; pressure inside falls below atmospheric pressure and air enters.

Gaseous exchange in other organisms

1 Large number – a large surface area; gill filaments overlap – slowing water flow – allowing more time for gaseous exchange; thin barrier of two cell layers – a short diffusion pathway; gill filaments have many blood capillaries.
2 Blood and water flow in opposite directions (countercurrent system); maintaining high concentration/ diffusion gradient.
3 Mouth closed and the floor of the buccal cavity is raised – reducing volume; increases pressure in buccal cavity – forcing water over gills through gill slits, into opercular cavity; pressure in opercular cavity rises – water forced out of opercular valve.
4 Carbon dioxide diffuses through stomata into intercellular spaces of the mesophyll; dissolves in moist cell walls, diffuses through cell wall and membrane into the cytoplasm.

Enzymes

1 Enzyme has a specific tertiary structure – determines shape of the active site; lipids shape to bind at the active site of lipases but carbohydrates have a different shape and cannot bind.
2 Enzyme concentration; substrate concentration; temperature; pH.
3 Denaturation – hydrogen and ionic bonds break – leading to irreversible change in tertiary structure; shape of the active site changes and substrate can't bind.
4 Add more substrate – if competitive inhibitor, rate of reaction increases; if non-competitive inhibitor, no change.

Digestion

1 Digestion breaks down large, insoluble compounds into small, soluble, easily absorbed molecules that can cross cell membranes.
2 Place equal sized mycelial discs of each fungus on separate starch-agar plates; leave at same temperature for 24 hours, then flood plates with iodine solution; compare size of clear zones – representing starch digestion and amylase activity.
3 Peristalsis – contraction of circular muscle and relaxation of longitudinal muscles behind move contents along the gut.

Digestion in humans

1 Amylase and maltase.
2 Emulsifies lipids to small droplets – providing large surface area for lipase to act on; bile is alkaline – providing optimum pH for pancreatic enzymes.
3

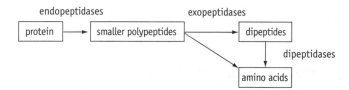

4 Large surface area – villi and microvilli; single layer of epithelial cells – short diffusion pathway; moist lining – soluble compounds dissolve for absorption; extensive blood capillaries – carry away absorbed digested products, maintaining high diffusion gradient; lacteals in villi absorb lipids – maintaining high diffusion gradient; mitochondria – ATP for active transport; carrier proteins in cell surface membranes.

Module 2: Genes and Genetic Engineering

Structure of nucleic acids

1 Pentose, phosphoric acid and an organic base.
2 Adenine, thymine, cytosine and guanine.
3 By hydrogen bonds between base pairs: A–T, C–G.
4 (i) RNA – single strand, DNA – double strand; RNA – ribose, DNA – deoxyribose; RNA – base uracil, DNA – thymine; (ii) tRNA is 'clover leaf' in shape; tRNA has an amino acid binding site; tRNA has anticodon, mRNA has codons.

DNA replication and the genetic code

1 When DNA replicates, each new DNA molecule has one original polynucleotide strand.
2 DNA polymerase joins nucleotides to form complementary strand to the original DNA strand.

3 An allele is one form of a gene.

4 A gene is the sequence of DNA nucleotide bases coding for the production of a specific polypeptide.

5 Some amino acids are coded for by more than one base triplet.

DNA and protein synthesis

1 A A C G C U G C A C C G U

2 DNA replication involves two template strands, transcription involves one; replication – DNA polymerase, transcription – RNA polymerase; replication uses thymine, transcription uses uracil.

3 Protein/polypeptide.

4 Each type carries a specific amino acid to the ribosomes. Anticodon on tRNA binds to codon on mRNA.

Gene mutation

1 High energy radiation, e.g. X-rays, gamma rays, UV light; high energy particles, e.g. alpha and beta particles; chemicals such as benzene.

2 Due to degeneracy of the DNA code the same amino acid may be coded for / one amino acid changed might not significantly affect tertiary structure.

3 Frame shift occurs – changes the sequence of amino acids from the point of the deletion – changes tertiary structure.

Mitosis

1 For growth and asexual reproduction.

2 Represents the number of paired chromosomes in a normal body cell. In man this is 46, i.e. 23 homologous pairs.

3 (i) Anaphase; (ii) prophase; (iii) interphase.

Meiosis and sexual reproduction

1 Female gametes are larger – more cytoplasm and food reserves; male gametes usually produced in much larger numbers; male gametes are mobile – swim to female gamete using a 'tail'.

2 A reproductive cell – a sperm or ovum.

3 In diploid organisms, meiosis leads to haploid gametes, which fuse at fertilisation to form a zygote, restoring the diploid number; ensuring that each generation inherits a constant number of chromosomes.

Genetic engineering

1 (i) Restriction endonuclease enzyme cuts DNA at a specific base sequence to cut donor or vector DNA;

(ii) ligase enzyme joins together plasmid DNA and 'foreign' DNA (gene) at 'sticky ends'.

2 Plasmids are small, circular sections of DNA found in some bacteria, often used as vectors in genetic engineering.

3 Genetic markers identify successfully genetically engineered bacteria by enabling them to survive exposure to specific chemicals which destroy non-recombinant bacteria.

4 Enzymes, hormones e.g. insulin, antibiotics.

Genetically modified animals

1 Organisms that have a gene from a different species.

2 Alpha-1-antitrypsin.

3 Transfer of foreign genes to non-target organisms; unknown ecological and evolutionary consequences; development of resistant species, e.g. antibiotic resistance bacteria; accidental transfer of unwanted genes by vector, e.g. virus.

Cloning

1 (i) A group of genetically identical organisms produced from one parent by asexual reproduction involving mitosis; (ii) a small piece of growing (meristematic) plant tissue (e.g. shoot tip) cut from a stock plant.

2 Sterilisation so no unwanted microorganisms enter culture.

3 Immature embryo → identical
twins → clones → implantation

Gene therapy and cystic fibrosis

1 Treatment of a disease by introducing copies of a healthy gene to replace the function of a defective gene.

2 Cystic fibrosis transmembrane regulator protein.

3 Accumulation of thick mucus narrows air passages, restricting air flow to the lungs – causing breathing difficulties.

4 Liposomes and viruses.

DNA technology

1 A primer molecule begins the formation of a complementary DNA strand.

2 16.

3 They stop DNA replication at the point they are used.

4 DNA fragments are separated according to size by gel electrophoresis and position of newly formed DNA fragments is identified by autoradiography.

Module 3(a): Physiology and Transport

Transport systems – the heart

1 Atria contract, then the ventricles; one-way valves prevent backflow; atrioventricular, atria to ventricles; semilunar, arteries to ventricles.

2 Between points X and Y, 0.8 seconds – rate = 60 divided by 0.8 = 75 beats per minute.

3 For blood to get round all of the body it enters/leaves the heart twice – once to the lungs and then to the body.

Transport systems – blood vessels

1 Artery wall approximately twice thickness of vein.

2 Arteries and veins – walls many cells thick/too thick for rapid diffusion; capillary wall one (endothelial) cell thick – short diffusion pathway for rapid diffusion.

3 Pulse felt when artery wall bulges, due to increased blood pressure caused by each heartbeat; artery wall recoils due to elastic tissue and muscle in the artery wall. (Note: the pulse/bulge travels along an artery as the blood flows.)

Transport systems – exchange

1 Haemoglobin 63% saturated with oxygen at pH 7.4 – 45% saturated at pH 7.2 – a fall of 18%.

2 Carbon monoxide binds to haemoglobin rather than oxygen – less oxygen carried to muscles for respiration – exercise is harder.

3 Oxygen diffuses across walls of alveolus and capillary along a concentration gradient; binds to haemoglobin in red blood cells; blood carries red cells to the left atrium and then left ventricle of the heart; blood is pumped along the aorta towards the legs; in capillaries in muscles, oxygen dissociates from haemoglobin – diffuses across capillary wall into tissue fluid and into muscle cells along a concentration gradient.

Control of breathing

1 Lung inflation stimulates stretch receptors; nerve impulses to medulla inhibit the inhalation centre; leading to exhalation; stretch receptors no longer stimulated, inhalation centre is not inhibited – leading to inhalation.

2 Lactic acid lowers pH of the blood (like carbonic acid), stimulates chemoreceptors in aortic body/carotid body and medulla.

Control of heartbeat

1 Time delay means the atrium contracts before the ventricle, so blood flows into the ventricle.

2 If the heart beats too fast heart muscle might be fatigued/damaged, leading to death/heart attack/damage to the heart.

Energy sources

1 Before exercise, muscle uses fatty acids for respiration; switches to glucose when exercise starts and uses up glycogen reserves; then goes to anaerobic respiration – aerobic respiration is not fast enough and oxygen supply to the muscle is too slow.

2 Sprinter uses oxygen and glucose already in the muscle cells, plus anaerobic respiration as needed.

Transport in plant roots

1 A – endodermis; B – xylem; C – phloem.

2 Water continually moving up the plant; ions dissolved in the xylem; ions actively transported into xylem by endodermal cells.

3 Water moving through the interconnecting cytoplasm of cells via plasmodesmata.

4 Casparian strip prevents water moving through the apoplast – forces water to cross cell membrane into the symplast – allowing control of water movement.

Transport in the phloem

1 Sucrose.

2 Sugars and other photosynthetic products cannot pass the ringed area due to absence of phloem. The fruit acts as a the main sink and accumulates these compounds.

3 (i) Produces photosynthetic products; (ii) uses or stores photosynthetic products.

Transport in the xylem

1 Measure volume of sap from freshly cut root stump; lower the temperature/deprive the root of oxygen/add cyanide and see if the rate of production slows.

2 Solar energy.

3 Cohesion – attractive force between water molecules due to hydrogen bonding; adhesion – attraction between the water molecules and the xylem walls.

Transpiration

1 Air movements remove water vapour from the leaf surface; increasing the water potential gradient and rate of transpiration. Increase in temperature increases rate of transpiration – provides water molecules with more kinetic energy, allowing them to evaporate more readily.

2 Plants that live in habitats where water is in short supply; having structural adaptations that reduce the rate of transpiration.

3 Thickened waxy cuticle reduces evaporation; curled leaves reduce the surface area for evaporation; reducing the water potential gradient for water loss by hairs on leaf surface – trap a layer of air which becomes saturated with water vapour, increasing the humidity in air around the stomata – reducing transpiration; reduced leaf surface area over which transpiration can occur; sunken stomata in pits – become saturated with water vapour.

Answers to end-of-module questions

Module 1: Core Principles

1 a i Cellulose molecules are very long polysaccharides – their size makes them insoluble (and unreactive); being straight they lie closely side by side and bind by many hydrogen bonds – form strong fibres.

ii Starch molecules are very long, coiled and branched polysaccharides – can be hydrolysed to glucose for respiration; large – insoluble, not affecting water potentials – unable to cross cell membranes and stays in the cell; branched – a compact store of chemical energy.

b

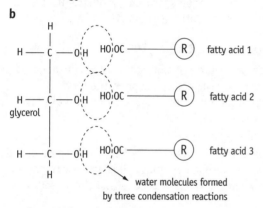

water molecules formed
by three condensation reactions

One mark each for structure of glycerol; fatty acid; and showing where water comes from.

c Primary structure – order of amino acids in a polypeptide determines whether the polypeptide forms a helix or folds on itself; secondary structure – where the chain folds and where bonds form between different parts of the chain; tertiary structure – gives a specific shape to the protein/polypeptide.

2 a i Sucrose – hydrolyse with acid, neutralise with alkali, test with Benedict's reagent, look for red precipitate.

ii Protein – Biuret test/copper sulphate solution and hydroxide, lilac colour is a positive result.

iii Lipid – add water and ethanol and shake, formation of an emulsion is a positive result.

b Digest protein with acid/protease, carry out chromatography, use two-way chromatography to separate amino acids that overlap with the first solvent, identify amino acids by R_f values.

3 a i Palisade cell has a cell wall, a large vacuole and chloroplasts.

ii Three of the following: cell membrane – controls movement of substances in and out of the cell; nucleus – contains the genetic information of the cell; mitochondrion – ATP is made in aerobic respiration; ribosomes – where proteins/polypeptides are made; rough ER – where proteins are made into the space of the ER; smooth ER – transports proteins made by the RER; Golgi body – packages proteins for transport round/out of the cell.

b Break open cells in ice-cold buffer; differential centrifugation to isolate cell fractions; electron microscopy to identify fraction with mitochondria; test mitochondrial fraction for its biochemical functions.

4 a i Arrows pointing from –350 kPa to –375 kPa and –400 kPa cells; also from –375 kPa to –400 kPa.

ii The more negative the water potential, the lower the concentration of water in the cell; water diffuses along concentration gradient from higher to lower; across the selectively permeable cell membranes; osmosis.

b i Chloride ions moved against concentration gradient; by active transport.

ii May be different numbers of carrier proteins for each ion; chloride ions might balance several types of positive ions.

iii Less oxygen – lower rate of respiration; less ATP for active transport; so internal concentrations of ions may fall.

5 a Movement of ions/molecules:

i randomly from higher to a lower concentration until evenly distributed;

ii by diffusion along a concentration gradient, through a membrane carrier or channel protein;

iii against a concentration gradient, using ATP from respiration.

b Water is a small, non-polar molecule which can pass through proteins in the phospholipid bilayer – glucose is much larger and has many polar groups, so cannot pass through bilayer – carrier protein binds to and carries glucose across.

6 a X has a greater surface area to volume ratio than Y; gives a greater surface area for the diffusion of gases; surface of Y is not large enough to exchange enough oxygen and carbon dioxide for all its tissues.

b Many stomata (so no leaf cell is far from one); leaf is thin in cross-section – short diffusion pathways; air spaces – rapid diffusion of gases; cell walls of mesophyll cells give large area for gas exchange.

c Both gas exchange surfaces have: a large surface area; a thin surface which is moist and has a good blood supply.

7 a Enzymes are very specific: react with a particular substrate; lower the activation energy of chemical reactions; allow reactions to occur at the temperatures found in living organisms.

b i To start with rate increases with the concentration of substrate because more active sites are occupied as the substrate concentration increases; graph levels off when all active sites are occupied most of the time.

　ii The inhibitor lowers the maximum rate of action of the enzyme; it is a non-competitive inhibitor; binding to a site other than the active site, changing the shape of the enzyme and the active site.

8 a 45°C.

b i Low kinetic energy, so few collisions between enzyme and substrate.

　ii Enzyme denatured; tertiary structure changed; enzyme no longer fits active site.

c i Non-boiled becomes active as kinetic energy increases; boiled, no activity because denatured.

　ii pH of the solution – at optimum concentration of substrate – higher produces higher rate; concentration of enzyme – if higher produces higher rate.

9 a Protein is digested by: acid hydrolysis in the stomach, to polypeptides; endopeptidase in the

stomach hydrolyses protein to smaller polypeptides; endopeptidase in the small intestine also produces smaller polypeptides; exopeptidases remove amino acids/dipeptides from 'ends' of polypeptides; endopeptidases produce 'ends' for exopeptidases – speeding digestion of protein; dipeptidases hydrolyse dipeptides to amino acids.

b i Oesophagus has well developed circular and longitudinal muscles; for peristalsis which pushes food towards the stomach; folded lining – expansion as food bolus passes/not adapted for absorption of food molecules.

　ii Stomach has muscles that churn contents; cells in gastric pits produce mucus to protect the stomach from acid and digestive enzymes; cells produce endopeptidases to digest protein.

　iii Small intestine has highly folded lining to increase the surface area for absorption; villi and micro-villi; muscle layers producing peristaltic movements that mix the gut contents and move them along; epithelial cells secreting digestive enzymes; and cells have carrier proteins to absorb specific substances and digestive enzymes 'built into' cell membranes.

c Slows digestion of fats; bile usually released in large volumes when fat present; emulsifies fats; giving larger surface area for action of lipases.

10 a Amylase in saliva; hydrolyses; some starch to maltose; amylase in small intestine hydrolyses starch to maltose; amylase from pancreatic juice; maltase hydrolyses maltose to glucose.

b Glucose can be absorbed by facilitated diffusion; and active transport; involving carrier proteins; which are part of the cell membranes; of epithelial cells.

Module 2: Genes and Genetic Engineering

1 a A = phosphate. B = pentose. C = thymine. D = cytosine. E = hydrogen bonds.

b Deoxyribose in DNA, ribose in RNA. Thymine in DNA, uracil in RNA. DNA is double-stranded, RNA is single stranded.

2 a i Transcription.

　ii Six-codon CGU appears twice.

　iii GCAT.

b i Anticodon.

　ii tRNA is 'clover leaf' in shape. tRNA standard length, mRNA is variable. tRNA has an amino acid binding site. tRNA has anticodon, mRNA has codons. tRNA has hydrogen bonds between base pairs.

　iii Carries specific amino acid to ribosome. Anticodon attaches to codon on mRNA; leaves amino acid at ribosome and picks up same type of amino acid from cytoplasm.

3 a DNA helix uncoils. Individual DNA nucleotides align according to specific base pairing. A with T, C with G. DNA polymerase joins nucleotides together. Hydrogen bonds between base pairs. Identical molecules produced, each with one original strand.

b High-energy radiation, e.g. X-rays. High-energy particles, e.g. alpha particles. Chemicals, e.g. benzene.

c Causes a change in the amino acid sequence of the protein produced. The protein is usually non-functioning and may affect the survival of the organism.

4 a 387.

b Change in base sequence of gene. This results in change in sequence of amino acids in the polypeptide/protein. Tertiary structure of enzyme is altered. Substrate cannot bind to active site.

c Deletion alters base sequence from point of mutation / causes a 'frame shift'. Sequence of amino acids is altered from this point. Therefore the protein is often non-functional. A single base substitution may not alter the amino acid coded for. Substitution only affects one amino acid. Consequently protein produced may still function.

5 a A = anaphase. B = interphase. C = metaphase.

b i $\frac{90}{275} \times 100 = 32.7$.

ii Oxygen availability. Temperature. Nutrient availability.

iii Replication of DNA. Protein synthesis. Synthesis of ATP.

6 a (i) 22. (ii) 44.

b Chromosomes carrying genes that control the same characteristic. The chromosomes are the same size.

c Meiosis reduces the diploid number to the haploid number. Gametes are haploid. Fusion of haploid gametes at fertilisation restores the diploid number.

d Splitting immature embryo provides identical cells. Process repeated several times. Developing clones are implanted into surrogate mothers.

7 a i Use of restriction endonuclease. Enzyme cuts DNA at specific base sequences.

ii Use of a vector e.g. plasmid. Cut plasmid with endonuclease. Join gene to DNA using ligase enzyme.

b Low temperature in North Sea. Slows down growth / activity / enzyme action of bacteria.

c Use of plasmid containing gene marker, e.g. for antibiotic resistance. Marker gene will enter the *super Pseudomonas* with foreign gene. Culture bacteria on plate with antibiotic. Only *super Pseudomonas* will grow.

8 a i Enzymes consist of protein. Enzymes denature at high temperatures.

ii Joins nucleotides together. According to base sequence on template strand.

b 256.

c Provide more sample for analysis from crime scene. Replicate trace DNA from extinct organisms. Provide copies of DNA for gene therapy.

9 a Antibiotics. Enzymes. Hormones.

b To prevent contamination by other organisms.

c Rapid rate of production due to fast growth rate of microorganisms. Less pollution / toxic products. Microorganisms can be genetically engineered. Less expensive running costs, e.g. raw products.

10 a CFTR controls movement of chloride ions in and out of cells. Defective protein prevents loss of chloride ions and accumulation of sodium ions in cell. Water remains in cell due to osmosis so that thick, sticky mucus is produced.

b Blocks pancreatic duct, reducing enzyme release. Slows down digestion. Mucus reduces efficiency of absorption.

c Normally functioning CFTR gene is cloned. Use of vector (viruses or liposomes) to transfer gene. Vector transfers gene to epithelial cells of lung. These cells secrete normal mucus.

Module 3(a): Physiology and Transport

1 a Atrioventricular valves; prevent backflow of blood from ventricle into atrium; semilunar valves; prevent backflow of blood into ventricles from arteries.

b i X – semilunar valve opens; because ventricle contracts – pressure inside becomes higher than in the aorta. Y – semilunar valve closes; because ventricle is relaxing – pressure inside falls below that in aorta.

ii Time between corresponding points on the graph = 0.8 seconds. Rate = 60/0.8 = 75 beats per minute.

2 a i Artery – thick muscular wall, to resist high blood pressure; narrow lumen, to maintain high pressure; lots of elastic tissue as well as muscle in its wall, to recoil from stretching caused by pulses of blood from the heart; this maintains high blood pressure; sheath of fibrous tissue, prevents splitting of the artery;

inner lining highly folded, allows expansion with each pulse of blood.

ii Capillary wall one (endothelial) cell thick; gives short diffusion pathways for exchanges between blood and tissues.

b Blood volume passing any point in a given time depends on rate of flow and cross-sectional area of the blood vessel. Arteries – very fast flow rate but a small cross-sectional area lumen. Veins – slow flow rate but very large cross-sectional area lumen.

3 a

Substance(s)	Site of capillaries where it enters the blood	Site of capillaries where it leaves the blood
Oxygen	**Alveoli**	Tissues of rest of body
Carbon dioxide	**Body tissues**	**Lungs**
Glucose, amino acids, fatty acids and mineral ions	Epithelium of villi of small intestine	**Body tissues**
Hormones	**Endocrine glands**	Target organs/tissues
Urea	**Liver**	Kidney

b i Loss of water to tissues – lowers hydrostatic pressure in capillary at venous end; makes water potential of remaining blood plasma more negative than tissue fluid; so water returns to the plasma by osmosis.

ii Lymph cannot drain back to the blood; tissue fluid accumulates in tissues; tissues swell.

4 a Sinoatrial node; in wall of right atrium; produces regular bursts of electrical impulses; spread rapidly through walls of both atria – they contract together; impulses reach the AV node; delay of 0.15 seconds – so that ventricles contract after atria; impulses from AV node through bundle of His and branches to all parts of the ventricles; which contract from bottom to top, pushing blood up and out into the arteries.

b i During exercise need to get more blood to muscles; carrying oxygen and glucose; for faster respiration.

ii Contracting muscles press on veins, force blood towards heart, causing greater filling of the ventricles; makes heart beat faster and stronger. Rises in carbon dioxide/fall in oxygen in blood stimulates chemoreceptors; carotid/aortic body; nerve impulses to cardioaccelerator centre; in medulla; nerve impulses to SAN.

5 a Chemoreceptors; aortic body/ carotid body/medulla; stimulated by lower pH; nerve impulses to medulla; breathing centre; increases in the rate and depth of breathing.

b Lactic acid lowers blood pH; lowers binding of oxygen to haemoglobin/shifts oxygen dissociation curve to right; more oxygen released to muscles.

6 a At rest muscle uses fatty acids for respiration; from triglycerides; use glucose during exercise; some from glycogen.

b Lactate diffuses out into the blood; travels to the liver; converted back to pyruvate; then glucose – returned to muscles or converted to glycogen.

7 a W – epidermis; X – endodermis; Y – phloem; Z – xylem.

b Water uptake mainly by root hairs; by osmosis; along water potential gradient; soil water higher water potential than root hairs; due to active uptake of mineral ions by hair cells; water moves along water potential gradient across cortex to the xylem vessels; via apoplast, symplast and vacuolar; enters symplast in endodermal cells; because of Casparian strips; water potential gradient is maintained by water continually moving up the xylem and by dissolved ions in the xylem sap.

8 a Cohesion–tension hypothesis; heat energy evaporates water from leaves; from mesophyll cells next to air spaces; water vapour diffuses out through the stomata into the air; transpiration; water potential of mesophyll cells lower than inner mesophyll cells; water moves by osmosis; along water potential gradient; by apoplast, symplast and vacuolar pathways; drawing water from xylem; creating a tension; "pulling up" water and dissolved ions; continuous water column from leaves to roots; maintained by cohesive forces between water molecules; adhesive forces between water molecules and xylem walls; due to hydrogen bonding.

b The following increase the rate of transpiration: a rise in temperature; a higher wind speed; a lower humidity. Temperature increases the rate of evaporation; higher wind and lower humidity both increase the water potential gradient for diffusion of water vapour.

9 a Mass flow hypothesis; translocation in phloem; sugars produced in mesophyll cells in leaves by photosynthesis/source; actively transported into sieve tubes; lowers their water potential so water enters by osmosis; creates high hydrostatic pressure in sieve tubes of leaf; in growing flower, sugars leave phloem to be used in respiration/growth/sinks; water potential of sieve tubes becomes higher and water leaves by osmosis; hydrostatic pressure falls and photosynthetic products are transported along pressure gradient in sieve tubes by mass flow.

b Ringing removed phloem; no translocation of sugars to roots; roots and trees die; no leaves to stop light getting to the ground; crops can grow between trees.

10 a (1) Active uptake into xylem; ringing leaves xylem in place; phosphate ions carried up in transpiration stream. (2) Ringing removes phloem; no mass flow from leaf source; so sugars stopped above ring.

b Autoradiography; X-ray film over plant; dark spots where radioactive tracers moved to.

c Respiration needed for active transport; radioactive carbon dioxide into sugars moved by mass flow; involving active transport of sugars into phloem; so cyanide inhibits sugar movement and radioactive carbon tracer; phosphate ions moved in transpiration stream; passive – due to evaporation of water from leaves; some effect due to role of active uptake of ions in roots.

Index